What Every Engineer Should Know About Excel

What Every Engineer Should Know: A Series

Series Editor*
Phillip A. Laplante
Pennsylvania State University

---

* Founding Series Editor: William H. Middendorf.

What Every Engineer Should Know About Excel

Second Edition

Jack P. Holman and Blake K. Holman

CRC Press
Taylor & Francis Group
Boca Raton London New York

CRC Press is an imprint of the
Taylor & Francis Group, an **informa** business

CRC Press
Taylor & Francis Group
6000 Broken Sound Parkway NW, Suite 300
Boca Raton, FL 33487-2742

International Standard Book Number-13: 978-1-138-30614-1 (Hardback)
International Standard Book Number-13: 978-1-138-03530-0 (Paperback)

Library of Congress Cataloging-in-Publication Data

Names: Holman, J. P. (Jack Philip), author.
Title: What every engineer should know about Excel / J.P. Holman and Blake K. Holman.
Description: Second edition. | Boca Raton : Taylor & Francis, CRC Press, 2018. | Series: What every engineer should know series | Includes bibliographical references and index.
Identifiers: LCCN 2017035509 | ISBN 9781138035300 (pbk.) | ISBN 9781138306141 (hardback) | ISBN 9781315268583 (ebook)
Subjects: LCSH: Microsoft Excel (Computer file) | Engineering--Computer programs.
Classification: LCC TA345 .H65 2018 | DDC 005.54--dc23
LC record available at https://lccn.loc.gov/2017035509

Visit the Taylor & Francis Web site at
http://www.taylorandfrancis.com

and the CRC Press Web site at
http://www.crcpress.com

Contents

Preface to the First Edition

This collection of materials involving operations in Microsoft Excel is intended primarily for engineers, although many of the displays and topics will be of interest to other readers as well. The procedures have been generated randomly as individual segments, which were distributed to classes as the need arose. They do not take the place of the many excellent books on the subjects of numerical methods, statistics, engineering analysis, or the information that is available through the Help features of the software packages. Some of the suggestions offered will be obvious to an experienced software user but will be less apparent or even eye-opening to others. It is this latter group for whom the collection was assembled.

Some of the materials were written for use in classes in engineering laboratory and heat-transfer subjects, so several of the examples are slanted in the direction of these applications. Even so, topics such as solutions to simultaneous linear and nonlinear equations and uses of graphing techniques are pervasive and easily extended to other applications.

The reader will notice that a basic familiarity with spreadsheets, the formats for entering equations, and a basic knowledge of graphs are assumed. A basic acquaintance with Microsoft Word is also expected, including simple editing operations.

The Table of Contents furnishes a fairly straightforward guide for selecting topics from the book. The topics are presented as stand-alone items in many cases, which do not necessarily depend on previous sections. Where previous topics are relevant, they are noted in that section. The reader will find that some topics are repeated—such as instructions for formatting graphs and charts—where it was judged beneficial.

In Chapter 1, the convention employed for sequential sets of operations is noted along with the background expected of the reader. The user will find the suggested custom keyboard setup in Section 2.3 to be very useful for typing equations and math symbols. While possibly of infrequent use, the application of photo inserts is discussed in Section 2.9. Increased use of scanners and digital cameras may add to the utility of these sections.

Most engineering graphs are of the x–y scatter variety, and the combination of the information presented in Section 3.3 and the suggested default settings in Section 3.22 should be quite helpful in applying these graphs. Most people do not think of using Excel to generate line drawings. The discussion in Section 4.2 illustrates the relative simplicity of making such drawings and embedding them in Excel and Word documents. Sections 4.3 and 4.5 illustrate methods for inserting and combining symbols, equations, and graphics in both Excel and Word.

Chapter 5 presents methods for solving single or simultaneous sets of linear or nonlinear equations. Section 5.4 presents an iterative method that is particularly useful for solving linear nodal equations in applications with sparse coefficient matrices. Histograms, cumulative frequency distributions, and normal probability functions are discussed in Chapter 6 along with several regression methods. Three regression techniques are applied to an example that analyzes the performance of a commercial air-conditioning unit.

Because financial analysis is frequently a part of engineering design, Chapter 7 presents an abbreviated view of the built-in Excel financial functions. Several examples of the use of these functions are also given. Optimization techniques are a part of engineering design; Chapter 8 gives a brief view of the use of the Solver feature of Excel for analyzing such problems.

Pivot tables are employed for arranging and categorizing small or large data sets into different formats. In Chapter 9, the approach has been to employ their use not only for rearranging tabular information but also for inserting calculated results of interest. This presentation then becomes a vehicle to supplement the creation of data tables and charts by other means.

Jack P. Holman

Preface to the Second Edition

The first edition of this book was written by my father, Dr. Jack P. Holman, noted engineering professor and author. His commitment to education was unequaled in his profession. Dr. Holman lived his life with high standards, high expectations, and a focus on continuous learning. This second edition extends Dr. Holman's initial work, updating it to the current version of Microsoft Excel (2016), and expands its scope to include data management, connectivity to external data sources, and integration with "the cloud" for optimal use of the Microsoft Excel product.

The advancement of Microsoft Excel since the first edition made several things either non-applicable or obsolete. These include the following that Dr. Holman called out in his preface to the first edition:

- Section 2.9 is no longer present, as Microsoft has a simple Insert Picture menu option available. Using the built-in Help function provides sufficient information on this operation.

- Section 3.22 of the first edition is no longer present in providing suggested default settings for line and scatter charts.

- In the first edition, there were significant differences between Excel and Word in their capabilities for use of symbols and equations. Since the introduction of the ribbon bar into the Microsoft Office products, Microsoft has made their products much more consistent in capability, including the use of symbols and equations. For this reason, this second edition has far less emphasis than the first edition on creating symbol or equation elements in Word and then transferring them to Excel.

With regard to the expanded scope of this edition, Chapter 10 provides several ways in which Excel can be interfaced to or integrated with external data sources for data management purposes. Chapter 11 provides a brief introduction to "cloud" services and capabilities where Excel can be very useful.

Blake K. Holman

About the Authors

Jack P. Holman received a PhD in mechanical engineering from Oklahoma State University (OSU). After research experience at the Air Force Aerospace Research Laboratories, he joined the faculty of Southern Methodist University, Dallas, Texas.

Dr. Holman has published over 30 papers in several areas of heat transfer and was the author of three widely used books: *Heat Transfer* (10th edition, 2009), *Thermodynamics* (4th edition, 1988), and *Experimental Methods for Engineers* (8th edition, 2012). These books have been translated into Spanish, Chinese, Japanese, Korean, Portuguese, Thai, and Indonesian and are distributed worldwide.

A fellow of ASME, Dr. Holman received numerous awards for his contributions to engineering and engineering education. He was awarded the Worcester Reed Warner Gold Medal and the James Harry Potter Gold Medal from ASME for distinguished contributions to the permanent literature of engineering. He received the American Society for Engineering Education's George Westinghouse and Ralph Coats Roe Awards for distinguished contributions to mechanical engineering education. In 1993, Dr. Holman was awarded the Melvin R. Lohmann Medal from OSU and was posthumously inducted into OSU's Engineering Hall of Fame in 2015.

Blake K. Holman received his bachelor's degree in mechanical engineering from Southern Methodist University and began his career in information technology as a management consultant in Dallas, Texas.

In 2005, Mr. Holman became chief information officer of Ryan, LLC, the world's largest independent tax consulting firm, based in Dallas, Texas. During his tenure at Ryan, *InformationWeek* magazine recognized his accomplishments, ranking Ryan number 130 on the InformationWeek 500 in 2011 and ranking Ryan number 98 on the InformationWeek 500 in 2012. In 2013, *ComputerWorld* magazine recognized Mr. Holman as one of their Premier 100 IT Leaders, a recognition that honors individuals who have had a positive impact on their organizations through technology.

Mr. Holman is currently the chief information officer of St. David's Foundation in Austin, Texas and was recognized in the Fall of 2016 as a finalist for Austin IT Executive of the Year.

1

Introduction

1.1 Getting the Most from Microsoft Excel

Microsoft Excel is a deceptive software package in that it offers computation and graphics capabilities far beyond what one would expect in a spreadsheet tool. Its capabilities remain unknown to many engineers and technical persons, although more engineers are adopting its use. This book is written for the person who is casually familiar with Excel but is unaware of its broad potential. Although a novice will find the material useful, it will be most attractive to those who have the following:

1. A basic knowledge of both Excel and Microsoft Word, including procedures for entering equations in Excel, editing fundamentals, and some experience with creating graphs
2. A basic knowledge of differential and integral calculus
3. For some sections, a familiarity with solution techniques for single and simultaneous equations
4. For some sections, a familiarity with basic statistics, including the concepts of standard deviation and probability

Many of the sections in this book resulted from small instructional sets that were written as stand-alone packages for engineering students enrolled in a mechanical engineering curriculum. In addition, some of the sets and example problems are related to applications in the thermal and fluids sciences areas of mechanical engineering. Although these application examples are retained, they are presented as part of more general procedures that are applicable to other engineering and technical disciplines.

Unless a person works with a software package such as Excel on a continual basis, it is easy to forget some of the shortcuts and nuances of operation that accomplish calculation or presentation objectives, namely procedures for viewing all equations on a worksheet, inserting symbols in equations, etc. Such hints are presented in compact form for the reader's convenience.

The title of this book refers to Excel, but the reader will find several applications that call for a combination of features of Word in conjunction with the capabilities of Excel. Microsoft PowerPoint is also a powerful tool for presentations but is not covered in this book.

The Help features of both Excel and Word are of obvious practical utility in working with the software. When appropriate, the reader's attention is directed to the Help facility for further information. Many books written on Excel and many specialized references pertain to particular engineering examples. A list of all references for this book is given in

1

the appendix, and callouts to this list are made at appropriate times in the book. Separate reference lists are not provided at the conclusion of each chapter.

Many worked examples are presented throughout the book. For the reader's convenience, each example is given a title. In some cases, the example title also specifies the calculation principle or technique that is being demonstrated. The book makes extensive use of graphs and figures, as well as printouts of specific spreadsheet segments employed in the examples. Screenshots that show worksheet and dialog window contents in perspective are also displayed.

The reader will find that many sections in the book can be used independently. This stand-alone nature results from the manner in which many of the topics were initially generated, as well as from an expectation that many readers want information in a compact self-contained form without having to move back and forth from section to section. To further achieve a compact presentation, explanatory notes are sometimes displayed as embedded text on the pertinent worksheet. When a topic relates to other sections, appropriate notes and references are given.

1.2 Conventions

As described earlier, many of the presentations are in a compact form, which allows for more rapid or convenient use. All references to Excel or Word in this book refer to the 2016 versions of these products.

When specifying a procedure that consists of a sequence of menu or ribbon bar operations, we will use the following convention

VIEW/SHOW/Gridlines

instead of the more cumbersome set of instructions:

1. Click the View tab of the ribbon bar
2. Go to the Show section of the ribbon
3. Click Gridlines

Another example in Excel is

INSERT/ILLUSTRATIONS/Pictures

which is equivalent to the following:

1. Click on the Insert tab of the ribbon bar
2. Go to the Illustrations section of the ribbon
3. Click on Pictures

Embedding of text boxes and descriptive statements in the example Excel worksheets is freely employed to express the instructions in a compact form. In many cases, this results in a font size smaller than that in the main body of the text, but is usually not objectionable.

1.3 Introduction to the Microsoft Office Ribbon Bar

Beginning with Office 2007, released in 2006, Microsoft introduced the ribbon bar for navigation within each Office application. In addition to being a new user interface, Microsoft introduced a level of consistency across the Office applications that was not previously seen.

The ribbon bar occupies a material amount of space on the top of the screen, but affords the user the opportunity to have many useful options available at the click of a button. The ribbon bar is organized into tabs, and each tab has several sections that present the various choices available under that tab. To insert a picture into a document, whether it be Word or Excel, the user must only click on the Insert tab of the ribbon bar, go to the Illustrations section of the ribbon and click on Pictures to accomplish their task. As noted in Section 1.2, we would note this operation as INSERT/ILLUSTRATIONS/Pictures.

If the reader has been using Office versions later than 2007, the ribbon bar should be familiar. For those who might still be using Office 2003 or earlier, they are encouraged to use an Internet search engine to obtain a more extensive introduction to the Office ribbon bar that allows them to navigate within the Microsoft Office applications.

1.4 Outline of Contents

Chapter 2 presents a potpourri of miscellaneous topics in Word and Excel that are applicable to the other chapters. Chapter 3 describes several graphing techniques that may be employed in engineering applications. Chapter 4 discusses the use of line drawings and other graphics in Word and Excel. Chapter 5 presents a variety of Excel techniques for solving single and multiple linear or nonlinear equations, along with numerical examples of each technique. Chapter 6 presents other numerical applications, including histograms and multivariable regression analysis, whereas Chapter 7 is devoted to the discussion and use of financial functions built into Excel. Chapter 8 presents optimization techniques that may be exploited with Excel Solver. Chapter 9 presents some basic, but very useful, operations with pivot tables. Chapter 10 presents techniques for interfacing Excel with external data sources and using such data in various operations. Finally, Chapter 11 presents an introduction to the use of Excel in "the cloud" as well as integration with other cloud applications that the engineer or technology professional might find useful.

2

Miscellaneous Operations in Excel and Word

2.1 Introduction

This chapter contains a collection of useful operations for editing or formatting, or simply a shortcut to doing something. The reader should take particular note of Section 2.3, which offers detailed suggestions for customizing the keyboard for direct typing of math and Greek symbols. Using these shortcuts will greatly simplify typing many equations and mathematical expressions, particularly if typing symbols and equations is a frequent task. For those that require significant entry of equations, Excel and Word have advanced equation-editing capabilities easily accessible from the Insert tab of the ribbon bar. Section 4.3 provides more in-depth explanation of this capability.

Some of the formatting and editing instructions are repeated from time to time when they are needed in a particular example or discussion.

2.2 Generating a Screenshot

It is often very useful to generate a picture of the contents (or a portion of the contents) of the computer screen at a particular moment in time. This can involve contents of the entire screen, a single window on the screen, or a subset of a window (or windows) on the screen.

Microsoft operating systems continue to support full screen capture and current window capture via the Print Screen and ALT+Print Screen key sequences. Using the Screen Clipping capability in Excel or Word, however, the user can get a full screen capture, a single window capture, or a large or small portion of the screen of the user's choice.

Before beginning the screen capture process, have a target document open and ready for insertion of the screen capture. The following are the instructions on using the Screen Clipping option and inserting the screen clipping into the target document, for each of the following three scenarios:

Capturing the entire screen

1. To generate a screenshot of the currently selected window on the screen, press the ALT+Print Screen keys.
2. Navigate to the document location where the screen capture will be inserted.
3. (a) Right click at the location and select from the Paste options available, or (b) press the CTRL+V keys to paste the screen capture at the current location.

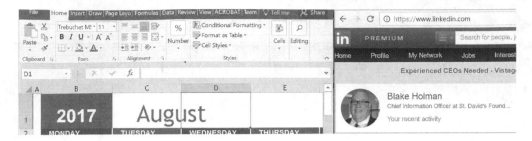

FIGURE 2.1

Capturing a single window on the screen

1. To generate a screenshot of the entire screen, press the Print Screen key.

Capturing a portion of the screen not bound by a window

1. To generate a screenshot of an unbounded portion of the screen, use the INSERT/ILLUSTRATIONS/Screenshot menu option and click Screen Clipping. Doing so will then minimize the program in which the user is inserting the screen clipping and will return him/her to what the screen looked like right before he/she started to perform the insert operation.

2. Select an area of the screen with the mouse—click to start the selection and then drag the mouse until the desired area is highlighted. Upon releasing the mouse button, the selected area will be inserted into the document at the active cursor location when the insert operation was initiated.

Figure 2.1 shows the result of a screen clipping that crosses multiple windows on the screen.

2.3 Custom Keyboard Setup for Symbols in Word or Excel

The following procedure may be used in either Word or Excel to customize individual keyboard keys for frequently used symbols:

1. Open new document.
2. Click INSERT/SYMBOLS/Symbol/More Symbols.
3. Select Font: Symbol or any other desired style.
4. Click on the desired symbol.
5. Click Shortcut key.
6. Press alternative keys or a combination of keys.
7. Click ASSIGN.
8. Click CLOSE.
9. Repeat this procedure to insert as many symbols and characters as desired.
10. Click Close to return to the document.

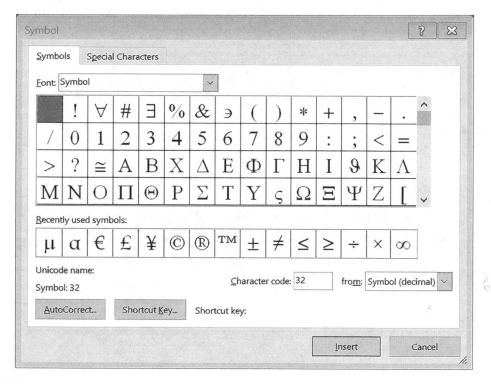

FIGURE 2.2

New shortcuts are saved in the default document template for Word or Excel so that they are available for all documents. For reference, the symbol font is shown in Figure 2.2, and a sample custom setup for shortcut keys is shown in Figure 2.3.

2.4 Viewing or Printing Column and Row Headings and Gridlines in Excel

To view or print column and row headings and gridlines:

1. Click PAGE LAYOUT/SHEET OPTIONS/Gridlines and check Print for printing of gridlines (Figure 2.4).
2. Click PAGE LAYOUT/SHEET OPTIONS/Headings and check Print for printing of headings (Figure 2.5).

2.5 Miscellaneous Useful Tips and Shortcuts

Through years of using Excel, many useful capabilities have emerged time and time again. The items listed down are a short list of these miscellaneous useful items—the reader is

','	6,°	d,δ	I,∫	O,Θ	S,Σ	y,∂
$,¢	7,→	D,Δ	k,κ	p,φ	t,θ	z,ξ
+,•	8,←	e,ε	l,λ	P,π	T,τ	
„,≤	9,↔	f,f	m,μ	q,⇒	v,ν	
.,≥	a,α	g,γ	n,η	r,ρ	w,ω	
=,±	b,β	G,Γ	N,≠	R,ℜ	x,≈	
/,÷	c,©	i,∞	o,Ω	s,σ	X,×	

FIGURE 2.3
Suggested custom shortcut key setup for Microsoft Word symbols: Type ALT+first symbol to get the symbol after the comma.

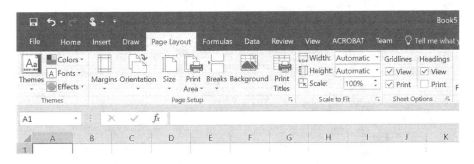

FIGURE 2.4

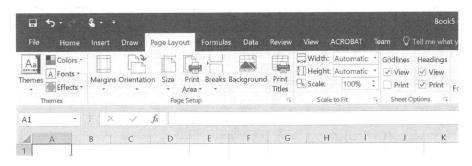

FIGURE 2.5

encouraged to discover his/her own shortcuts and add to them. Where seemingly appropriate, some of these are repeated in examples throughout this book.

Listing recently used Excel files

FILE/OPTIONS/Advanced/Display and then choose the number of recent workbooks to list.

Moving and sizing charts and text boxes on a worksheet

To move the entire chart or text box, activate the chart by clicking on the CHART AREA, not the PLOT AREA, and dragging it to the new location.

To resize the chart, activate the chart, click on the corners or side handles of the CHART AREA until a double arrow appears, then drag to the desired proportions.

Adding or removing fill to cells or text boxes

Activate the object or area, click on the HOME/FONT/Fill Color icon, and select the desired fill color, pattern, or No Fill.

Adding or removing a line border on a text box

Activate the object, click on DRAW/PENS/Color to select a visible line color for the text box.

Changing border or drawing line weights

Activate the object, click on DRAW/PENS/Thickness, and make a selection.

Editing chart elements

Activate the chart. Click the+icon to the right of the chart and select from Titles, Axes, Gridlines, Legend, Error Bars, Data Table, and Data Labels.

Displaying formulas in cells

Press the CTRL+` key sequence to toggle back and forth between displaying cell formulas and cell values.

Adding (or deleting) sheet and page numbers

Click on PAGE LAYOUT/PAGE SETUP/Full page setup options and select the Header/Footer tab of the dialog box. Set up your desired header and footer.

Printing portrait or landscape page orientation

Select PAGE LAYOUT/PAGE SETUP/Orientation and choose the desired format: Portrait or Landscape.

Deleting in Word

To delete the word behind the cursor: press CTRL+Backspace.
To delete the word after the cursor: press CTRL+Delete.

Subscripts and superscripts in Word

Subscript: CTRL+the equal sign (=)
Reverse subscript: CTRL+the equal sign (=)
Superscript: CTRL+the plus sign (+) (using the Shift key)
Reverse superscript: CTRL+the plus sign (+) (using the Shift key)

Protecting worksheets

To prevent accidental typing over formulas or objects in a worksheet, lock the material in place by clicking FILE/PROTECT WORKBOOK/Protect Current Sheet or HOME/CELLS/Format/Protect Sheet and select the desired protections for the workbook or worksheet.

To reverse the protection action, click FILE, and on the PROTECT WORKBOOK selection, click Unprotect in the lower right corner of the selection. Alternatively, you can select HOME/CELLS/Format/Unprotect Sheet.

2.6 Moving Objects in Small Increments (Nudging)

To move an object by small increments:

1. Select the object by clicking it.
2. Press the arrow keys to move object in the desired direction.
3. Hold down the CTRL key while pressing the arrow keys to move the object in one-pixel increments.

2.7 Formatting Objects in Word, Including Wrapping

Charts, graphs, drawing objects, pictures, and text boxes may all be copied to Word from other sources, namely, Excel, and then adjusted in size or position, or wrapped with text. The procedure for making these adjustments is as follows:

1. Activate an object, a chart, a drawing, or a picture by clicking it.
2. Click LAYOUT/ARRANGE/Wrap Text/More Layout Options. The dialog window will appear as in Figure 2.6.
3. Select the tab of interest. In Figure 2.7, the Size tab is shown, which may be used to adjust the size of the object.

2.8 Formatting Objects in Excel

Drawing objects and pictures may be altered in size in Excel by dragging the edges to the desired size or by first activating the object and then using the Picture Tools FORMAT tab of the ribbon bar. For pictures, the tab of Figure 2.8 will appear, which allows modification of the picture size (far right end of the ribbon bar) and a variety of other adjustments including Corrections, Color, Artistic Effects, Borders, Other Effects, and even Cropping.

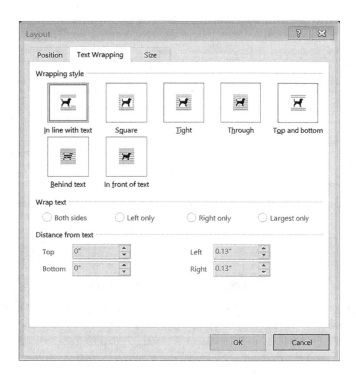

FIGURE 2.6

FIGURE 2.7

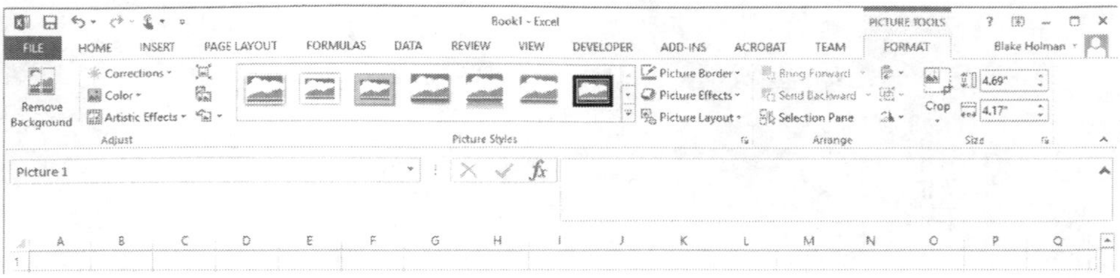

FIGURE 2.8

Recent versions of Excel have incorporated many functions that were previously present only in digital-photo-editing software.

2.9 Copying Formulas by Dragging the Fill Handle

Many engineering situations arise in which tabulations or plots of a function are needed for uniform increments in the argument of the function. This operation is very easy to perform in Excel by using the Fill Handle and dragging it. In Figures 2.9 through 2.11, we show how this is accomplished for the simple function $y = x^2$ in increments of $\Delta x = 0.1$ over the range $1 < x < 2$.

The start of the range for x is entered in cell A4 as 1. Then, the next value of x is entered in cell A5 as 1.1. Cells A4 and A5 are activated, producing what is shown in Figure 2.9 (note

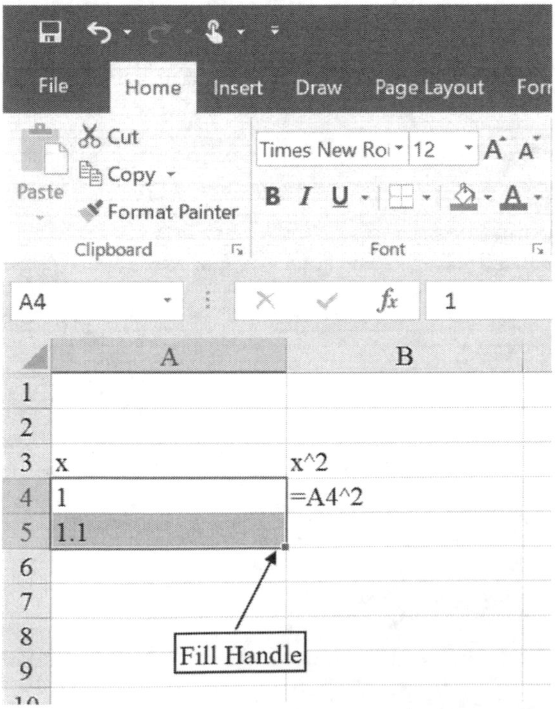

FIGURE 2.9

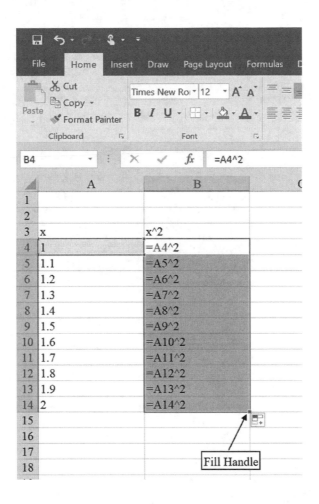

FIGURE 2.10

that CTRL+` was used to display the formulas). Then, the Fill Handle is clicked and dragged down for the desired number of increments, producing the result shown in Figure 2.10.

The formula for x^2 is entered in cell B4 as shown in Figure 2.9. This cell is activated and the Fill Handle is dragged down to copy the formula as shown in Figure 2.10. Toggling the formula view by pressing CTRL+` produces the final numerical results as shown in Figure 2.11. Display of the formulas is not necessary in the drag process, and the result in Figure 2.11 can be produced by drag-copying cell B4 while in the numerical display mode.

Copying of cell formulas could also be accomplished by activating the cell, clicking EDIT/COPY, and then dragging the mouse for the number of cells desired. Using the Fill Handle can be easier, though it is often a matter of personal preference.

2.10 Copying Cell Formulas: Effect of Relative and Absolute Addresses

Copying a cell formula can produce different results depending on whether absolute cell references are used or not. In cell B4 of Figure 2.12, the formula calls for the square of the value

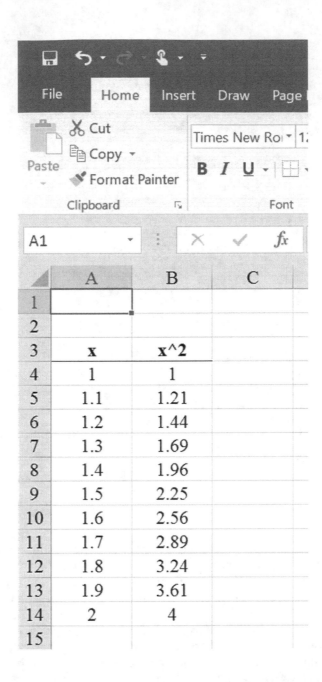

FIGURE 2.11

in cell F1. The same result is called for in the formula of cell C4. Using the cell reference $F$1 is an absolute cell reference to the value in cell F1, whereas using the cell reference F1 is called a relative cell reference. The results of copying these two formulas are shown on the worksheet. When B4 is copied to C8, the formula does not change because of the absolute cell reference $F$1. When C4 is copied, an entirely different set of results can be obtained as shown below:

1. When C4 is copied to D8, F1 becomes G5 (one column to the right—thus, F becomes G, and four rows down—thus, row 1 becomes row 5).

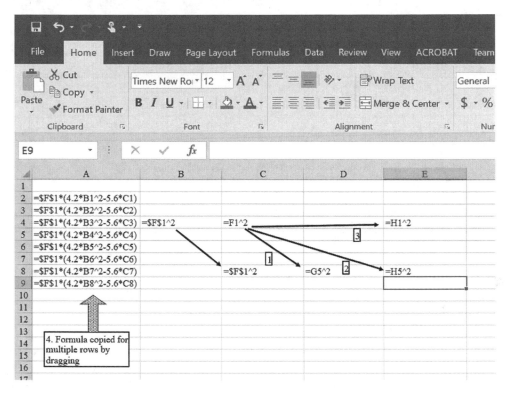

FIGURE 2.12

2. When C4 is copied to E8, F1 becomes H5 (two columns to the right—thus, F becomes H; and four rows down—thus, row 1 becomes row 5).

3. When C4 is copied to E4, F1 becomes H1 (two columns to right—thus, F becomes H; and the row remains the same, so the row number remains 1).

4. A formula may be copied for successive rows or columns as shown in column A. This is done by dragging the Fill Handle of a selected formula cell, a procedure outlined in Section 2.9. Note how the formula retains the absolute reference but changes the relative cell locations.

Note that moving a formula does not change the cell addresses in the formula. See "Moving Formulas" in Excel Help for details.

2.11 Shortcut for Changing the Status of Cell Addresses

The F4 key may be used to quickly change the absolute or relative status of a cell address. The procedure as applied to the formula in cell B4 of Figure 2.9 is as follows:

1. Activate the cell B4 containing the formula.
2. Highlight the A4 cell reference in the formula.

3. Press the F4 key until the desired type of cell reference is obtained. Repeated pressing of the F4 key will cycle through the four possible cell references as A4, $A$4, A$4, and $A$4.

4. Press Enter.

2.12 Switching and Copying Columns or Rows, and Changing Rows to Columns or Columns to Rows

Sometimes the position of data in a column or row needs to be switched in order to provide for a different orientation on a chart. When using x–y scatter graphs (Section 3.3), Excel treats the left column or the top row of data as the x or abscissa coordinate. The position of the column on the worksheet may be changed by copying one of the columns (or rows) to a new location by the following procedure:

1. Select (activate) the columns or rows of cells to be copied.

2. Click HOME/CLIPBOARD/Copy or press CTRL+C.

3. Click the cell that will be the top cell of the new column or the left cell of the new row.

4. Click HOME/CLIPBOARD/Paste. The menu shown in Figure 2.13 will appear. Under Paste, choose one of the Values options if new formulas are not to be created. See the earlier discussion on relative and absolute cell locations.

5. If a column is to be switched to a row or if a row is to be switched to a column, click Transpose.

6. Click OK.

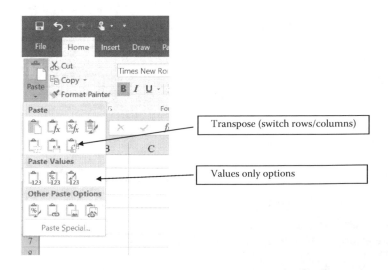

FIGURE 2.13

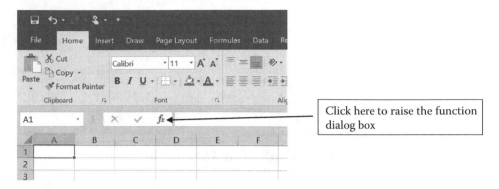

FIGURE 2.14

2.13 Built-In Functions in Excel

Excel has hundreds of built-in functions that may be accessed by the function name followed by the syntax that applies to that function. The reader who needs to apply these functions in worksheet formulas will usually be aware of the abbreviations assigned to the functions.

For a listing of functions, the reader may consult Excel Help by pressing F1 for additional details by entering search terms such as the following:

Engineering functions

Math and trigonometry functions

Statistical functions

Financial functions (for business users)

Alternatively, the reader may click on the function symbol on the formula bar (Figure 2.14), which will raise the dialog box shown in Figure 2.15. Selecting a category of functions from the drop-down menu will provide details on the functions available in that category.

For later reference, the user may wish to print out the list of functions. A complete description of each function can be called up by the function name using Help, which will display all syntax requirements. Some examples are given in Table 2.1. Financial functions are discussed in Chapter 7.

2.14 Creating Single-Variable Tables

Copying formulas in successive cells is one way to create a data table as described in Section 2.9. An alternative, and sometimes simpler, procedure uses the DATA/FORECAST/What-If-Analysis/Data Table command according to the following steps:

1. Set aside rows or columns in a worksheet for labeling variables.

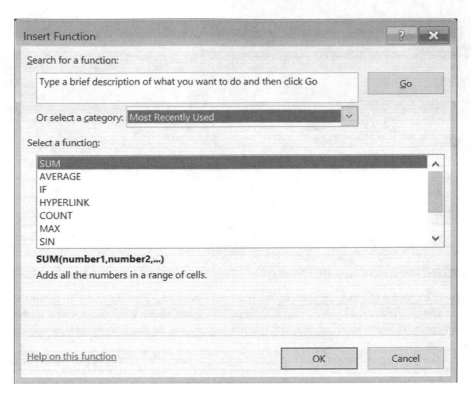

FIGURE 2.15

2. Choose a column to contain the numerical values of the input variables. Insert input values in this column. Increments may be set as described in Section 2.9 or by direct entry.

3. Type the formula to be calculated in the column to the right of the column in step 2 and one row above. The formula should be written in terms of an *input cell* that is located apart from the body of cells that will house the final table. Selection of the input cell is rather arbitrary. The only requirement is that it must be located outside the cell range assigned for the table.

4. Select (activate) cells containing values of the input variable, formula to be evaluated, and cells that will contain the results.

5. Click DATA/FORECAST/What-If-Analysis/Data Table.

6. Enter the input cell location for a column table in the dialog window.

7. Click OK. The table will appear.

8. If additional result functions need to be evaluated, enter the formulas for each in the cells adjacent to the formulas in step 3, and repeat steps 4 through 7.

9. The procedure may also be executed using rows for data input. In this case, the formulas are typed in the column to the left of the initial value and one cell below.

TABLE 2.1

Abbreviated List of Excel Built-In Functions

Function	Syntax	Arguments
Absolute value	ABS(x)	x=real number
Arccosine	ACOS(x)	$-1 < x < +1$, returns $-\pi/2$ to $\pi/2$
Hyperbolic arccosine	ACOSH(x)	x=number
Arcsine	ASIN(x)	$-1 < x < +1$, returns $-\pi/2$ to $\pi/2$
Hyperbolic arcsine	ASINH(x)	x=number
Arctangent	ATAN(x)	x=number, returns $-\pi/2$ to $\pi/2$
Hyperbolic arctangent	ATANH(x)	$-1 < x < +1$
Bessel function $J_n(x)$	BESSELJ(n, x)	n=order (integer), x=number
Bessel function $Y_n(x)$	BESSELY(n, x)	n=order (integer), x=number
Cosine	COS(x)	x=angle in radians
Hyperbolic cosine	COSH(x)	x=number
Error function	ERF(x)	$x \le 0$, returns value of 0–1.0
Exponential	EXP(x)	x=number, returns e^x
Natural logarithm	LN(x)	$x > 0$
Logarithm to base b	LOG(x, b)	$x > 0$, b=base (default b=10)
Logarithm to base 10	LOG10(x)	$x > 0$
Matrix inversion	MINVERSE	See Section 5.5
Matrix multiplication	MMULT	See Section 5.5
Pi	PI()	Returns numerical value of π
Sine	SIN(x)	x=angle in radians
Hyperbolic sine	SINH(x)	x=number
Square root (positive)	SQRT(x)	$x \le 0$
Square root of Pi	SQRTPI	Returns $\pi^{1/2}$
Summation	SUM($x_1, x_2, \ldots$ 255 values)	Sum of 255 values or array
Sum of squares	SUMSQ($x_1, x_2, \ldots$ 255 values, or array)	Sum of squares of 255 values or array
Tangent	TAN(x)	x=angle in radians
Hyperbolic tangent	TANH(x)	x=number
Arithmetic average	AVERAGE($x_1, x_2, \ldots$ 255 values)	Average of 255 values or array
Sum of squares of deviations from arithmetic mean	DEVSQ($x_1, x_2, \ldots$ 255 values, or array)	$= \Sigma(x_i - x_{mean})^2$ x_{mean}=arithmetic mean
Maximum, median, or minimum	MAX(), MEDIAN() or MIN($x_1, x_2, \ldots$ 255 values)	Returns values for 255 values or array
Normal distributions	NORMDIST, NORMINV, NORMSDIST, NORMSINV	See Section 6.5
R-squared	RSQ	See Section 3.8
Sample standard deviation	STDEV($x_1, x_2, \ldots$ 255 values)	Returns sample standard deviation of 255 values or array
Population standard deviation	STDweEVP($x_1, x_2, \ldots$ 255 values)	Returns population standard deviation of 255 values or array
Financial functions		See Chapter 7

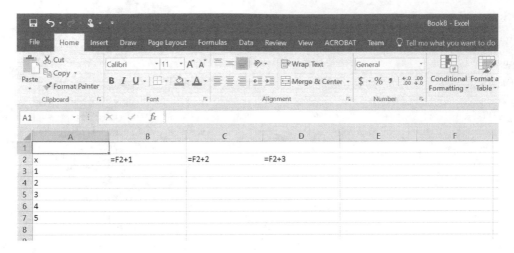

FIGURE 2.16

Example 2.1: Constructijon of a Table for Simple Functions of a Single Variable

We will construct a table for the following three functions of x over the range $0 < x < 5$ in increments of 1.0:

$$y_1 = x + 1$$

$$y_2 = x + 2$$

$$y_3 = x + 3$$

The worksheet is shown in Figure 2.16. Cell A2 is used for the x label. The three formulas for the functions are listed in cells B2, C2, and D2, respectively, and the cell range to house the table is A2:D7. An input cell apart from this region is chosen as F2 and the formulas written in terms of this cell are shown in Figure 2.16. Note that the CTRL+` key sequence was used to switch to formula viewing.

The table area is selected, the DATA/FORECAST/What-If-Analysis/Data Table menu option is clicked and Input column cell F2 is inserted in this dialog box, producing the result shown in Figure 2.17. OK is clicked. The result (formula view) is shown in Figure 2.18 while the numerical result is shown in Figure 2.19.

2.15 Creating Two-Variable Tables

Two-variable tables may be constructed using a procedure similar to that employed for one-variable tables. Two examples of formulas involving two input variables are:

$$z = (x_2 + y_2)^{1/2}$$

and

$$z = (x+1)(y+2)$$

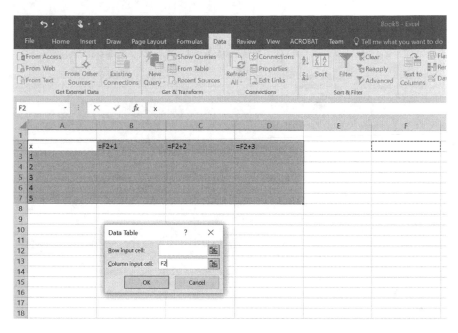

FIGURE 2.17

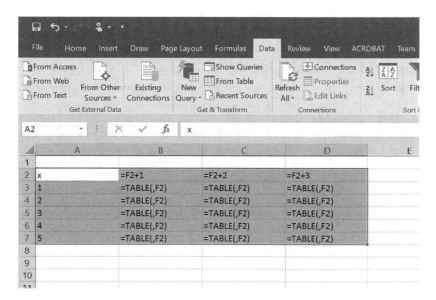

FIGURE 2.18

The procedure for creating the two-variable data table is as follows:

1. Select two input cells apart from the block of cells that will house the data table. These cells will serve as the variables in the formulas.

2. Choose a cell on the worksheet and enter the formula for the function in terms of the two input cells.

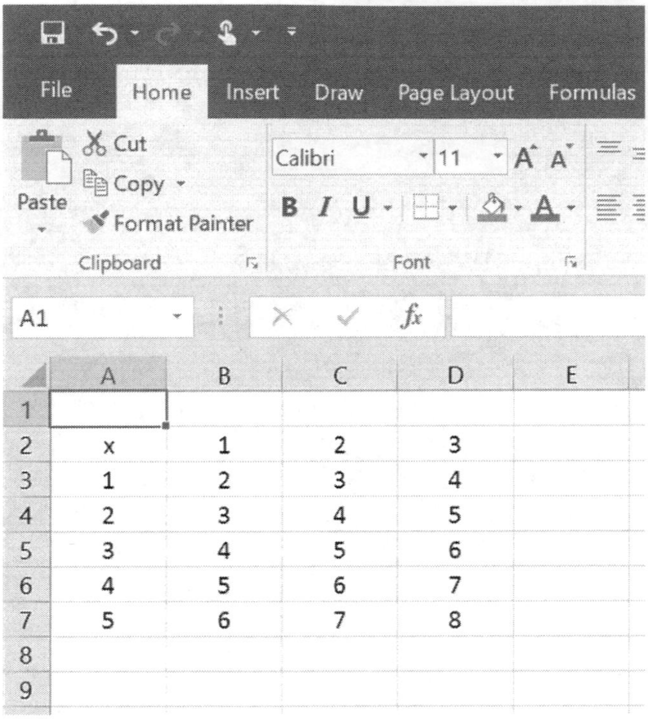

FIGURE 2.19

3. Enter a list of input values for one variable in the same column as the formula, but below the formula.

4. Enter a list of input values for the second variable in the same row containing the formula, but to the right of the formula.

5. Select (click and drag) the range of cells that are to contain the formula, input values of both variables, and data table.

6. Click DATA/FORECAST/What-If-Analysis/Data Table.

7. The dialog window will appear. Enter the row and column input cells used in writing the formula in step 2 and those corresponding to the input values entered in steps 3 and 4.

8. Click OK. The table will appear.

Example 2.2: Two-Variable Data Table

To illustrate this method, we will construct a data table for the function:

$$z = (x^2 + y^2) \text{ for } 1 < x < 5 \text{ and } 1 < y < 5$$

Increments of x and y are chosen as 1.0. The worksheet is set up so that cells H2 and I2 are chosen as input cells for x and y, respectively, and the formula for z is written in cell A2 as shown in Figure 2.20. The A column is chosen for x, with the five input values entered. Likewise, row 2 is chosen for y, with five corresponding input values.

Smaller or larger increments in x and y could be chosen and entered either directly or as described in Section 2.9.

Next, the table range A2:F7 is selected by click-dragging. DATA/FORECAST/What-If-Analysis/Data Table is clicked and I2 entered as the input cell for y along with cell H2 as the input cell for x. The entries are shown in the window of Figure 2.21. OK is clicked and the data table appears as shown in Figure 2.22, with the formulas displayed. Removing the formulas gives the final table shown in Figure 2.23.

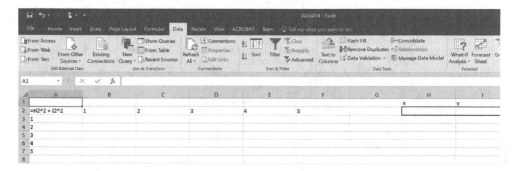

FIGURE 2.20

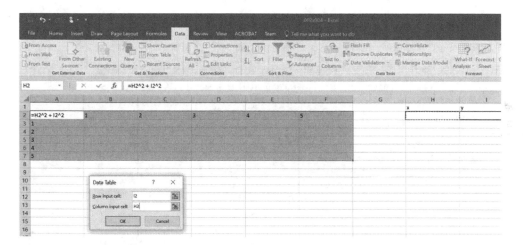

FIGURE 2.21

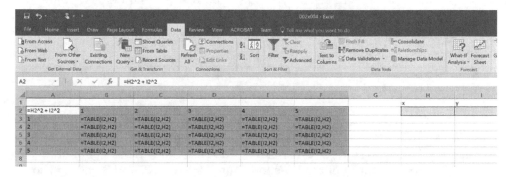

FIGURE 2.22

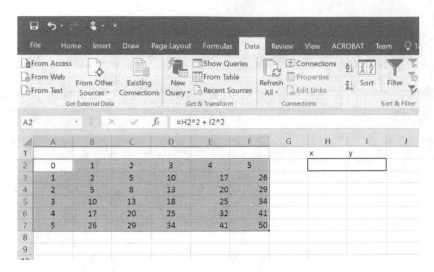

FIGURE 2.23

Problems

2.1 In Excel, click FILE/OPTIONS. Copy a portion of the Options window using the Screen clipping capabilities. Adjust the size of the inserted screen clipping. Move the window to new positions by pressing the cursor arrows or by dragging the image.

2.2 Customize the keyboard in Word or Excel as shown in Section 2.3 and type the following equations:

$$A = x_0 / \left\{ \left[1 - \left(\omega / \omega_n \right)^2 \right] + \left[2 \left(\omega / \omega_n \right) \left(c / c_c \right) \right]^2 \right\}^{1/2}$$

$$\theta / \theta^\infty = e^{-(hA/\rho cV)\tau}$$

2.3 Open a new Excel worksheet. Type a few comments or equations. Change the font for the worksheet to a different type and size to suit your personal interests.

2.4 Perform the drag-copying process as described in Sections 2.9 and 2.10.

2.5 Open an Excel worksheet and evaluate the following functions:

$e^{-0.5}$

cosh (2.3)

$Tanh^{-1}(0.5)$

Numerical value of π

2.6 Using the DATA/FORECAST/What-If-Analysis/Data Table command, construct a table of values of the function $\sin(nx)$ for $n=1, 2$, and 3 and $x=1$ to 1.5. Choose appropriate increments in x for the calculations.

2.7 Using the DATA/FORECAST/What-If-Analysis/Data Table command, construct a table of the three functions:

$$y = x^{1/2}$$

$$y = x + 0.3$$

$$y = x^2$$

over the range $0 < x < 5$.

2.8 Using the DATA/FORECAST/What-If-Analysis/Data Table command, construct a table of the Bessel function $J(n, x)$ for $n = 1, 2$, and 3 and $0 < x < 3$. Choose increments in x as desired.

2.9 Using the HOME/CLIPBOARD/Copy command, transpose the x–y column data in columns A and B into the row data shown:

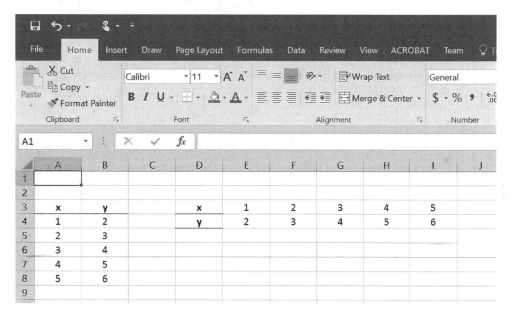

2.10 Enter the following values in an Excel worksheet:

1, 1.2, 1.1, 1.05, 0.96, 0.95, 1.06, 1.15, 1.21, 0.94, 1.01

and using built-in functions, evaluate:

$$y = \left\{ \left[\sum (x - x_m)^2 \right] / n \right\}^{1/2},$$

where

$$x_m = (\sum x) / n \text{ and } n = \text{number of values}$$

2.11 Compare the result of Problem 2.10 and the application of the worksheet functions STDEV and STDEVP to the data points.

3

Charts and Graphs

3.1 Introduction

The preparation, publication, and presentation of graphs and charts represent a significant portion of engineering practice. In Excel, a majority of such displays are given the designation of x–y scatter graphs. For this reason, we will concentrate our discussion on that type of graphical presentation. Bar graphs and column graphs are discussed briefly in Section 3.18, and surface (3-D) charts are discussed in Section 3.20. Obviously, the interested reader may explore other graphical possibilities.

The display and discussion in Section 3.3 categorize the five types of scatter graphs available in Excel, along with a general statement of an application for each type. Examples of data presentations using scatter charts are given in this chapter as well as in the application sections of other chapters. Treatment of math and other symbols in graphical displays is discussed in this chapter and in sections of Chapter 4 connected with embedded drawing objects. An important part of the present chapter is concerned with the display and correlation of data using trend lines and the built-in least-squares analysis features of Excel. Examples are given for correlation equations using linear, power, and exponential functions. Section 3.19 discusses formatting and cosmetic adjustments that are available for the various graphs.

As in other chapters in this book, many of the sections of this chapter are essentially self-contained and can be studied on a stand-alone basis. To provide for this capability, charts in some sections have been embedded with text along with a reduction in type size. As appropriate, cross-references are made to related sections of this and other chapters.

3.2 Moving Dialog Windows

A small data set is shown in Figure 3.1a. INSERT/CHART/Line is clicked, producing the Chart insertion shown in Figure 3.1b. If needed, the chart may be moved by clicking on the chart border and dragging it to a new position as shown in Figure 3.1c.

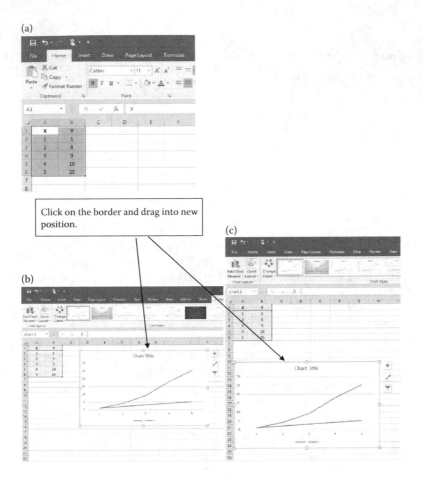

FIGURE 3.1

3.3 Excel Choices of x–y Scatter Charts

Excel offers several variations of x–y scatter charts. Figure 3.2 shows these options as a result of clicking on the INSERT/CHART/Scatter/More Scatter Charts … menu option. Each chart is discussed briefly below.

1. *Scatter*: Data plotted using data markers but no connecting line segments. This type of plot is employed for experimental data with considerable scatter but may be fitted with a computed trend line.

2. *Scatter with Smooth Lines and Markers*: Data plotted using data markers connected by smoothed lines as determined by the computer. This type of plot is employed for either calculated points or experimental data with rather smooth variations from point to point.

3. *Scatter with Smooth Lines*: Data plotted as in item 2 but without data markers. This type of plot is most frequently employed for calculated curves and is almost never used for presenting experimental data because the data points are not displayed.

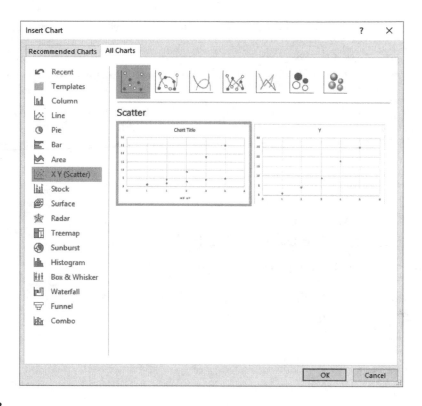

FIGURE 3.2

4. *Scatter with Straight Lines and Markers*: Data points plotted with markers and with points connected by straight line segments. This type of plot is sometimes employed for calibration curves in which linear interpolation between data points is assumed.

5. *Scatter with Straight Lines*: Data points plotted as in item 4, but without data markers. This type of plot is frequently used when points are obtained from a numerical analysis that assumes linear behavior between calculated points.

6. *Bubble*: Data points plotted as large circles, or "bubbles." This type of plot is useful when there is a need to show the relative size of a plotted value in relation to other plotted values.

7. *3-D Bubble*: Data points plotted as large spheres, or 3-D "bubbles." This type of plot is similar to a bubble plot when only the plotted points appear in three dimensions instead of two.

3.4 Selecting and Adding Data for x–y Scatter Charts

In setting up scatter charts, the x-axis will be either the left column or the top row of data, depending on whether columns or rows are chosen for the data series. The y-axis will be the remaining columns or rows. After the chart is established, the addition of data will be as a new y-axis regardless of their location relative to the column or row taken as the x-axis. The data selection procedure is as follows:

1. Click-drag cells for the x-axis and release the mouse button when the x-axis selection is complete.
2. Press the CTRL key and move the pointer to the start of the first y-axis data. Click-drag cells for the first y-axis data while holding down the CTRL key.
3. Continue this procedure for successive y-axis data, still holding down the CTRL key.

3.5 Changing/Replacing Data for Charts

The data for charts can be changed or replaced as follows:

1. Activate the chart. The Chart Tools menu options will appear on the Ribbon Bar.
2. Click CHART TOOLS/DESIGN/DATA/Select Data. A Select Data Source popup window will appear as in Figure 3.3, allowing you to expand or edit the existing data series.
3. In the Chart Data Range selection, either type or select a new range of data for the chart. The new range can include the old data range, or it can be a completely new data selection.
    a. For selecting new data, click the collapse button at the right end of the Chart Data Range field, and then proceed to select the worksheet data desired.

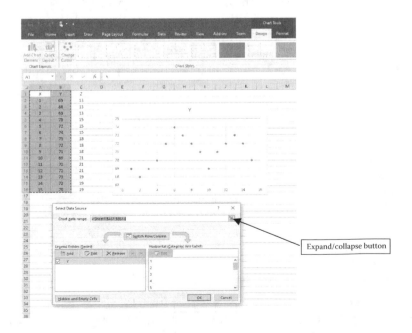

FIGURE 3.3

b. Select the *replacement* data cells to be added as described in Section 3.4. These replacement cells may be chosen to include or omit the old data cells. To add a data series while retaining the old data, see Section 3.6.

4. Click the collapse (expand) button again and the Select Data Source dialog box will reappear.

5. Click OK, which will redraw the chart with the new replacement data.

6. Make cosmetic and other adjustments to the chart as needed.

3.6 Adding Data to Charts

Data can be added to charts as follows:

1. Activate the chart. The Chart Tools menu options will appear on the Ribbon Bar.

2. Click CHART TOOLS/DESIGN/DATA/Select Data. A Select Data Source popup window will appear as in Figure 3.3. The lower left area of the Select Data Source popup window is where you will click Add to add another data series using the Edit Series dialog box that appears as shown in Figure 3.4.

3. In the Edit Series dialog box, select a name for the data series. Use the collapse/expand button if needed.

4. After selecting a name for the data series, select the Series X and Y values, again, using the collapse/expand button as needed.

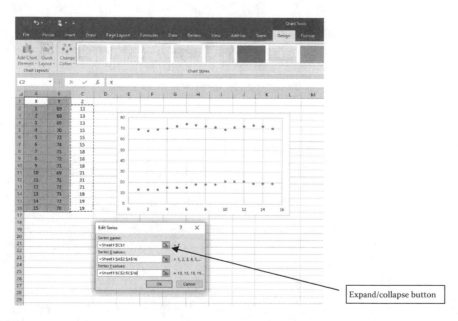

FIGURE 3.4

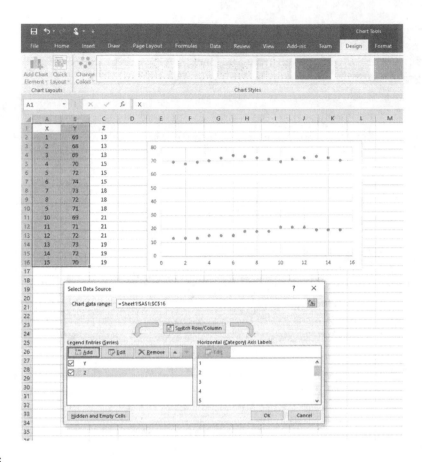

FIGURE 3.5

5. After completing the selections, click OK. The Select Data Source popup will reappear with the additional data series listed, and the chart will be updated with the additional data series plotted in the chart as shown in Figure 3.5.

6. Make cosmetic and other adjustments to the chart as needed.

3.7 Adding Trend Lines and Correlation Equations to Scatter Charts

To add a trend line, first click on the chart to activate it. In the upper-right corner of the chart, click the "+" sign to add a chart element. In the popup, select Trendline for a standard linear trend line, or select the right arrow to pick a type of trend line (Linear, Exponential, Linear Forecast, or Two Period Moving Average). Upon selecting the type, a prompt for additional information will appear. In the case of a simple x–y linear trend line, the prompt asks for the desired axis for the trend line—x or y.

When a trend line is added to a chart, Excel automatically calculates its R^2 value. To display an equation for the trend line and its R^2 value, double click the trend line on the chart to bring up a Format Trendline window. Toward the bottom of that window, there are two choices: Display Equation on chart and Display R-squared value on chart. Select from those two choices to display the items. See Sections 3.9 and 3.10 for specific examples.

3.8 Equation for R^2

The equation employed by Excel for calculating R^2 in the trend line fits is given by

$$R^2 = \left[n\Sigma x_i y_i - (\Sigma x_i)(\Sigma y_i) \right]^2 / \left[n\Sigma x_i^2 - (\Sigma x_i)^2 \right] \left[n\Sigma y_i^2 - (\Sigma y_i)^2 \right] \qquad (3.1)$$

R^2 is called the coefficient of determination, whereas R is called the correlation coefficient. This equation expresses what is called the Pearson correlation coefficient, which is demonstrated by the PEARSON worksheet function. A calculation of R^2 separate from the trend line determinations may also be obtained by calling either the worksheet function RSQ or PEARSON. Use the Excel Help facility for the proper syntax of these functions. The R^2 displayed with the graphical trend line is expressed as follows:

$$R^2 = 1 - \frac{SSE}{SST}$$

where SSE is the sum of the squares of the error from the correlating trend line, or

$$SSE = \Sigma(y_i - y_{ic})^2$$

and SST is the sum of squares of deviations from the arithmetic mean, $y_{mean} = (\Sigma y_i)/n$, and may be expressed in the form:

$$SST = \left(\Sigma y_i^2\right) - \left(\Sigma y_i\right)^2 / n$$

where y_{ic} represents the value of y on the linear trend line fit. For a perfect match between the data points y_i and the trend line, $R^2 = 1.0$. For exponential, power, and polynomial trend lines, Excel uses a transformed regression model. Note that these calculations are equivalent to using a population standard deviation instead of a sample standard deviation. Still, a perfect fit will be obtained when $y_i = y_{ic}$. SST may also be calculated in terms of the population standard deviation function STDEVP through the relation:

$$SST = n \times [STDEVP(y_i)]^2$$

3.9 Correlation of Experimental Data with Power Relation

Several physical phenomena follow a power law relation between variables. Examples are as follows:

$$Nu = C Re^n$$

for forced convection and

$$Nu = C(GrPr)^m$$

for free convection heat transfer. The general power law relation has the form

$$y = ax^b \qquad\qquad (3.2)$$

Taking the logarithm of both sides of the equation gives

$$\log y = \log a + b \log x \qquad\qquad (3.3)$$

which is a linear relation between log y and log x. When x and y are plotted on a log–log graph, b will be the slope of the line and log a will be the intercept at x = 1.0 (see Section 3.10). When trying to fit the experimental data with the power law relation, scatter in the data will normally occur and a least-squares analysis should be employed to determine the best fit. A correlation coefficient may also be calculated to indicate the goodness of fit.

Excel may be used to (1) display the data on a log–log plot, (2) calculate the values of the constants a and b using a least-squares analysis, (3) display the resultant correlation trend line, and (4) display the correlation equations on the plot.

Performing these steps in Excel, the procedure is as follows:

1. List the data in two columns. Label columns as appropriate. Consider discarding any data points that appear to be in gross error. This step may be deferred until after the data plot is obtained. See step 7.

2. Select the data to be plotted.

3. Click INSERT/CHARTS/Scatter and select the scatter chart *without* connecting the line segments (type 1 chart).

4. Click the chart to be edited. Double click either the x- or y-value axis—a FORMAT AXIS window will appear on the right side of the Excel worksheet. Under AXIS OPTIONS, select the upper and lower bounds for the axis as well as the major and minor units on the axis scale. Click Logarithmic Scale and the desired base (10 is default). Repeat for the other value axis. If desired, expand the TICK MARKS, LABELS, and NUMBER options and select the desired options for each axis.

5. Once step 4 is completed, click the chart again. Then, click CHART TOOLS/ DESIGN/CHART LAYOUTS/Add Chart Element/Trendline/More Trendline Options. Under Trendline Options, select Power, and click Display Equation on the chart and Display R-squared value on the chart. The chart will automatically update with the trend line, the equation, and the R^2 value.

6. Inspect the final graph. Does the trend line appear to represent the data? If not, the power relation may not be correct for the physical application. *This step is important! A correlation equation should NEVER be accepted without visual confirmation of agreement with the experimental data points.* The computer will perform the trend line analysis as instructed, but it cannot assure that the functional form selected is correct.

7. Examine the individual data points in the final plot. If some points appear to be widely scattered from the main body of data, consult the original data sheets for possible errors or erratic behavior in the experiment. Consider eliminating suspicious points.

8. If a decision is made to eliminate points as discussed in step 7, delete the respective entries in the data cells. The deletions will appear on the chart, and a new trend line and correlation equation will be displayed, based on the remaining data points.

9. Make final adjustments to the cosmetics of the chart, fonts, titles, etc. If a large number of data points are involved, some adjustments in the size of data markers or in the line width for the trend line may be in order.

Two examples of power law correlation plots are shown in Figure 3.6. One has a rather good fit, whereas the other has a lot of scatter. In the latter case, one should suspect that either the data are bad or that a power law relation does not fit the physical situation.

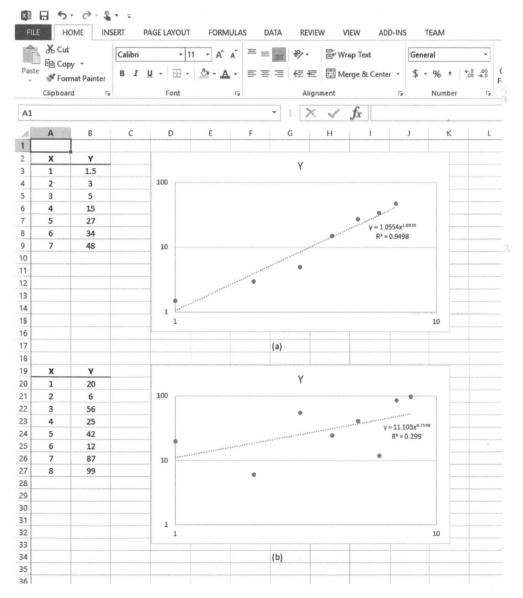

FIGURE 3.6

3.10 Use of Logarithmic Scales

The data are first plotted on a linear graph as shown in Figure 3.7a, indicating a decaying exponential or inverse power relation. Logarithmic scales are then selected by double clicking on each value axis. A FORMAT AXIS window will appear on the right side of the Excel worksheet. Under AXIS OPTIONS, select the upper and lower bounds for the axis as well as the major and minor units on the axis scale. Click Logarithmic Scale and the desired base (10 is default). Remember to repeat for the other value axis. For the y-axis, in the FORMAT AXIS section labeled "Horizontal axis crosses," set the "Axis value" field to 0.1 (the lower edge of the graph), and the result is shown in Figure 3.7b. Next, a trend line is added by clicking on the chart. Then, click CHART TOOLS/DESIGN/CHART LAYOUTS/ Add Chart Element/Trendline/More Trendline Options. Under Trendline Options, select Power, and click Display Equation on the chart and Display R-squared value on the chart. The chart will automatically update with the trend line, the equation, and the R^2 value. The result is shown in Figure 3.7c. Visual inspection indicates that a power relation does indeed fit the data.

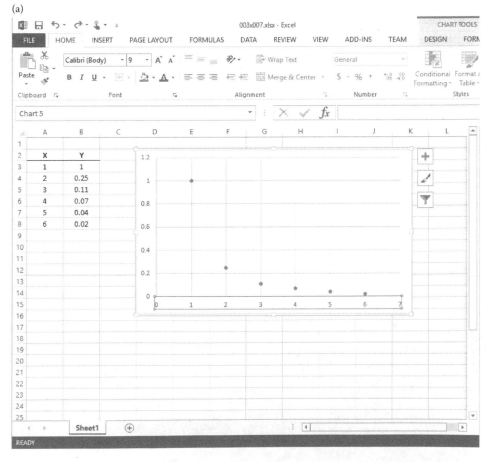

FIGURE 3.7

(Continued)

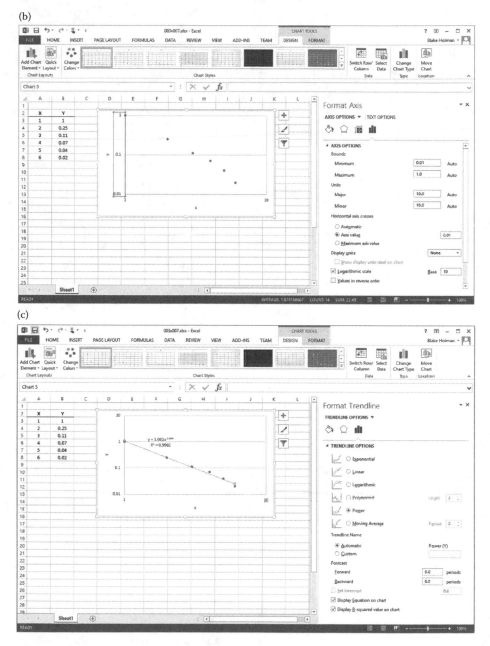

FIGURE 3.7 (CONTINUED)

3.11 Correlation with Exponential Functions

The exponential function $y = e^{-0.1x}$ ($y = EXP(-0.01x)$ as an Excel function) is tabulated and is shown first as a linear plot in Figure 3.8a with a linear trend line fit, which obviously does not fit. Second, a linear plot with an exponential trend line fit is shown in Figure 3.8b with perfect correlation. Third, the function is plotted on a semi-log graph that displays the

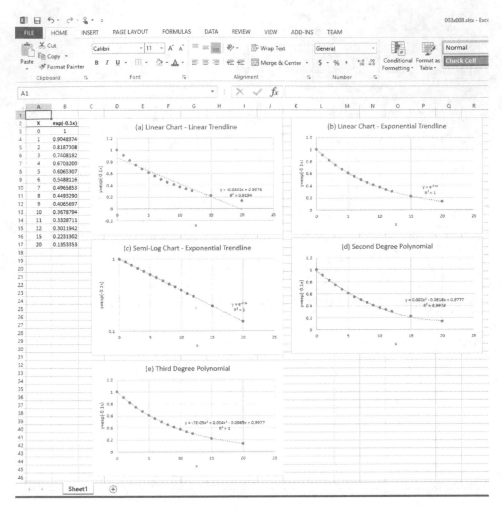

FIGURE 3.8

function as a straight line in Figure 3.8c. Again, an exponential trend line is fitted with perfect correlation. Inspection of the visual display is needed to evaluate the trend line fit. For comparison, the final two plots of Figure 3.8d and e show fits of second- and third-degree polynomials. The third-degree polynomial shows a perfect correlation. Polynomials may frequently be employed to obtain a good fit when the functional form is uncertain.

3.12 Use of Different Scatter Graphs for the Same Data

Figure 3.9 shows six scatter plots of a set of hypothetical experimental data displayed in the upper-left corner of the sheet. Figure 3.9a is a type 1 scatter graph, Figure 3.9b is a type 4 chart, and Figure 3.9c is a type 3 chart—all plotted with linear scales on both axes (see Section 3.3). Figure 3.9d through f are the same types of plots, but with logarithmic scales on the axes (see Section 3.10). The graph in Figure 3.9d shows that the data fall approximately

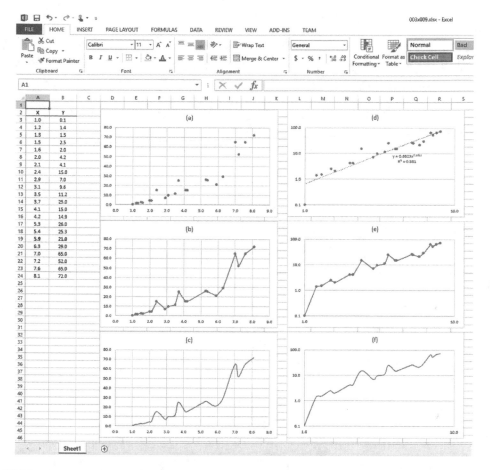

FIGURE 3.9

on a straight line so a power law relation might be anticipated. Inserting a trend line and the correlation equation and value of R^2 (see Section 3.9) provides confirmation of such.

Inspecting the data plot in Figure 3.10a, five data points seem out of place and are, hence, suspect. Four of those data points are circled in Figure 3.10a and one is circled in Figure 3.10d. Figure 3.10b and c are type 4 and type 3 scatter charts respectively for the data in Figure 3.10a. Figure 3.10e and f are also type 4 and type 3 scatter charts, but with logarithmic scales on the axes. If these points are eliminated as shown in Figure 3.11, a better correlation results. Similar to Figures 3.9 and 3.10, Figure 3.11 provides type 1, type 4 and type 3 scatter charts in Figure 3.11a through c respectively, and type 1, type 4 and type 3 scatter charts plotted on a logarithmic scale for Figure 3.11d through f respectively.

3.12.1 Observations

The charts in Figure 3.9c and f do not convey much information about the data and do not give the reader any hint of what might be going on with the experiment. Looking at the other charts would certainly not give one the impression of a smooth variation of y as a function of x. The charts in Figure 3.9b and e are better, but those in Figure 3.9a and d give the best impression of the scatter of data. The chart in Figure 3.9d, because it indicates that the data are approximately on a straight line in a log–log plot, gives the clue that a power

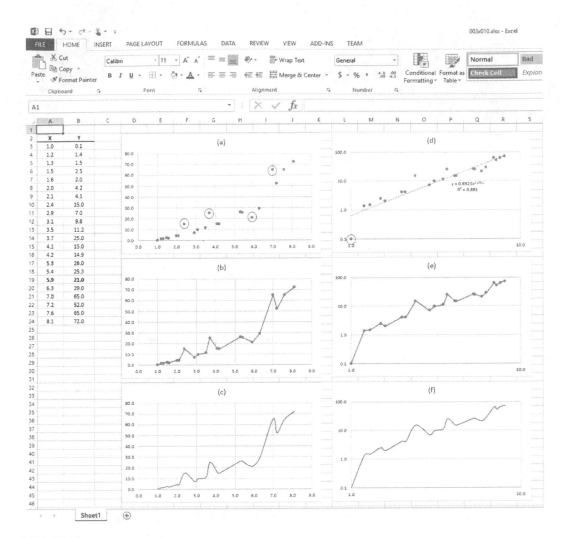

FIGURE 3.10

relation may apply if one deletes the first data point, which appears completely skewed. As we have stated before, one should never leave out the data markers when plotting experimental results. In other chart examples, involving plots of calculated points, we will see that the use of smooth curves as in Figure 3.11c and f will be appropriate.

3.13 Plot of a Function of Two Variables with Different Chart Types

This example illustrates how it is possible to present the plot of a function or data in different chart types to convey different impressions of the function. The Bessel function $J_n(x)$ is chosen for presentation because of its attractive appearance as a damped sine wave. The function is callable in Excel as BESSELJ(x, n). The worksheet is set up as shown in Figure 3.12, with column A listing the values of the argument x to be incremented using

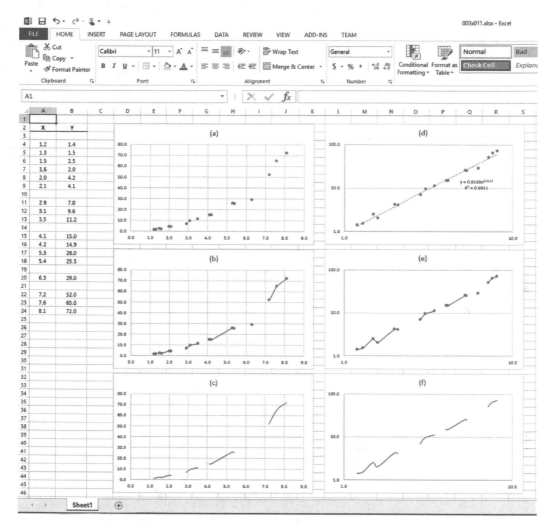

FIGURE 3.11

the Dx value in cell H3. These increments may be selected as coarse or as fine as desired. Columns B through F compute the Bessel functions as a function of the argument x and orders n = 0–4. The formulas are copied for as many rows as needed for the plot. In Figure 3.12, the two views of a brief worksheet are shown, one with formulas displayed (using CTRL+˘) and one with values displayed.

The different types of charts selected for presentation of the Bessel function are shown in Figure 3.13a through f. The chart in Figure 3.13a is a typical type 3 scatter graph with smooth curves connecting the points and no data markers. The chart in Figure 3.13b is an area chart showing the curves as overlapping with a shading effect. The chart in Figure 3.13c is a surface chart with a wire frame 3-D surface.

The charts in Figure 3.13d through f are all variations of the chart in Figure 3.13b. Starting with that chart, change the chart type via the CHART TOOLS/DESIGN/TYPE/Change Chart Type menu and select Area/3-D Area. To rotate the chart plot area, double click the plot area for the chart and go to the Plot Area Options. Under 3-D Rotation, enter a value

(a)

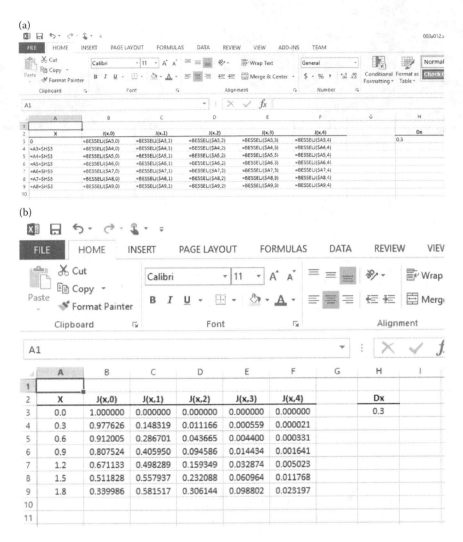

(b)

FIGURE 3.12

for X-Rotation. In the case of Figure 3.13d through f, rotation values of 289°, 329°, and 239° are chosen. The results are shown in Figure 3.13d through f. This presentation enables one to look at the "front" or "back" of functions or data. Other effects, such as y-axis rotation and perspective, are also available from the Plot Area Options panel.

3.13.1 Changes in Gap Width on 3-D Displays

The width of the separation gap between the plotted data series in a 3-D chart may be adjusted using the following procedure:

1. Activate the data series by clicking on it.
2. Click CHART TOOLS/FORMAT/CURRENT DATA SELECTION/Format Selection to bring up and make changes in gap width as desired. If multiple series are displayed, adjust the gap width for each data series in the chart.

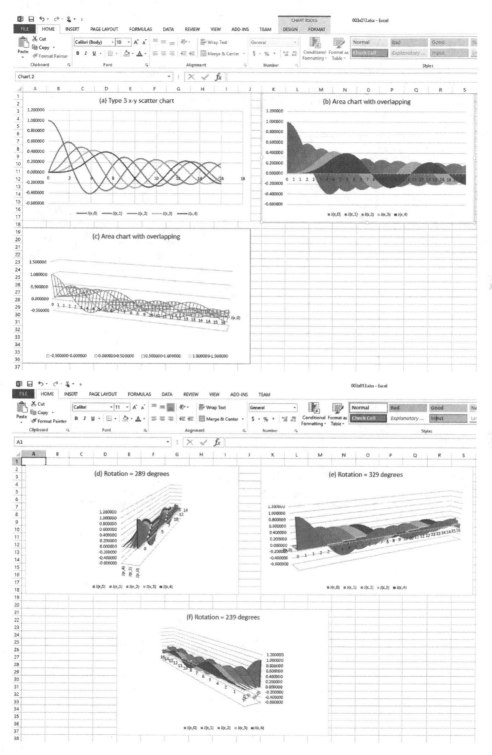

FIGURE 3.13

3.14 Plots of Two Variables with and without Separate Scales

Two sets of data (curves) with either markedly different ranges or units may be plotted as two data series on the same scatter graph. Both the abscissas (x-coordinate) and ordinates (y-coordinate) may have different scales or units. The procedure shown in Figure 3.14 is as follows:

1. Plot both sets of points using the normal procedure for scatter graphs, as shown in Figure 3.14a.

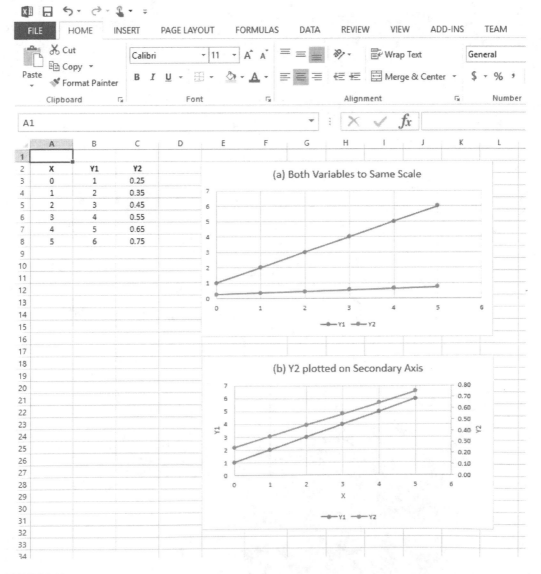

FIGURE 3.14

2. Double click on the Y2 data series to cause the Format Data Series window to appear. Click on Secondary Axis to create a secondary ordinate axis for Y2. The scale of the graph for that series will be expanded or contracted and the data will be replotted accordingly, as shown in Figure 3.14b.

3. Attach titles (labels) of indicated variables and units, tick marks, titles, etc., as appropriate. The data sets may be marked or titled using a separate legend box, different color lines, or box labels inserted directly on the chart itself.

Other cosmetic features may be added as necessary.

3.15 Charts Used for Calculation Purposes or G&A Format

Figure 3.15 shows a type 3 x–y scatter chart for displaying computed values of R, the capital recovery factor used in financial calculations. The Excel table of values is shown in the upperpart of the worksheet, followed by an equation for R. The use of smoothed curves without data markers is an excellent choice for the presentation in this example.

3.15.1 G&A Chart

In Figure 3.16, the same information is plotted in what we choose to call a G&A Chart (for Generals and Admirals). Still, a type 3 scatter chart is employed, but larger fonts are used for axis and chart labels where possible. Minor gridlines are deleted, and a light pattern is added to the body of the chart for cosmetic effects. This might be called a "broad brush" chart as it shows main trends. It should not be used for calculation purposes. The chart in Figure 3.15 can be used to read rather precise values of R.

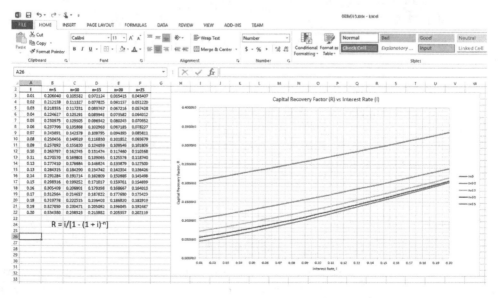

FIGURE 3.15

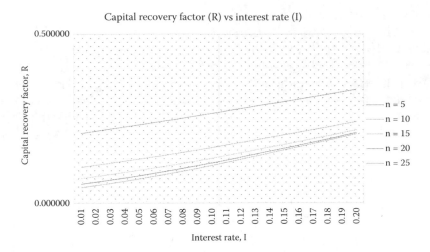

Capital recovery factor (R) vs interest rate (I)

FIGURE 3.16

3.16 Stretching Out a Chart

A chart that needs to extend broadly from top to bottom or side to side, but which appears compressed on a single page, can be stretched out or expanded as follows:

1. Click the chart to activate it.
2. If the chart is needed on a page by itself, open a new worksheet in the current Excel workbook.
3. Click the upper-left corner cell (A1) of the new worksheet or the cell at which the upper-left corner of the chart should be located.
4. Click EDIT/PASTE or CTRL+V.
5. The chart can be resized by using one of the following techniques:
   a. To resize the chart without maintaining the aspect ratio (ratio of length to width), click on a side or a corner of the chart and use the mouse to move that side or corner to the desired dimension.
   b. To resize the chart while maintaining the aspect ratio (ratio of length to width), click on a corner while holding down the SHIFT key and drag the corner to the desired size of the chart. The length and width will automatically adjust to maintain the aspect ratio.

3.17 Calculation and Graphing of Moving Averages

Moving averages are employed as forecasting tools in applications ranging from stock market predictions to estimations of sales and inventory trends. The calculation assumes that a forecast value of the variable under consideration may be made as a simple arithmetic average of the preceding actual values over a selected number of time periods. The number of periods is chosen to fit the situation. In many cases, moving averages are charted using several calculation intervals to gain comparative insights into the specific trends.

The formula for the moving average calculation is:

$$F_t = (1/n) \sum_{i=1}^{n} A_{t-i} \tag{3.4}$$

or

$$F_{t+1} = (1/n) \sum_{i=1}^{n} A_{t+1-i} \tag{3.5}$$

where

F_t = forecast value of the variable at time t

n = number of previous time periods over which the average is to be computed (Excel uses a default value of three periods if some other number is not specified)

A_t = actual value of the variable at time t

Thus, for n = 4 time intervals, we would have forecast values at times t = 6 and 7 of

$F_6 = (A_5 + A_4 + A_3 + A_2)/4$

$F_7 = (A_6 + A_5 + A_4 + A_3)/4$

Excel performs the calculation for a set of specified A_t values and presents a graph of the forecast values F_t along with the actual values for comparison. It is easy to change the number of periods for the moving average calculation to examine the influence of this selection on the forecasting trends.

Example 3.1: Weather Temperature Trends

Figure 3.17 displays three types of weather temperature data as indicated in the nomenclature for the figure: (1) TV fifth-day future forecasts for high and low temperatures, (2) actual high and low temperatures, and (3) long-term average or normal high and low temperatures. We will present the results of moving average calculations for 10-, 30-, and 60-day intervals over a 220-day total time period. The calculations will be made for the following:

1. The TV fifth-day future forecast for daily high temperature in °F
2. The long-term average high temperature in °F

In Excel 2016, plotting a moving average on a chart for a set of data is as simple as plotting a trend line, such as linear, power, and polynomial trend lines previously discussed.

Beginning with a chart containing a plot of the TV fifth-day future forecast for daily high temperatures, a moving average trend line is added by clicking on CHART TOOLS/DESIGN/CHART LAYOUTS/Add Chart Element/More Trendline Options. The Format Trendline Options window appears on the right side of the worksheet. From this window, we select Moving average with a Period value of 10, indicating a 10-day moving average. The result of this addition is shown in Figure 3.18.

Following the same steps, we add a 30-day moving average and a 60-day moving average to the same chart, which results in the chart shown in Figure 3.19.

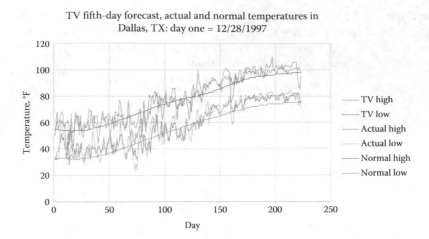

FIGURE 3.17

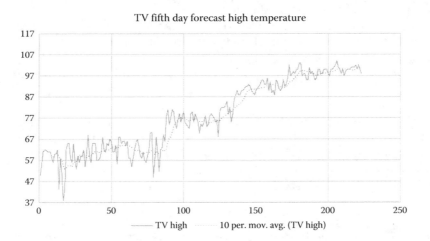

FIGURE 3.18

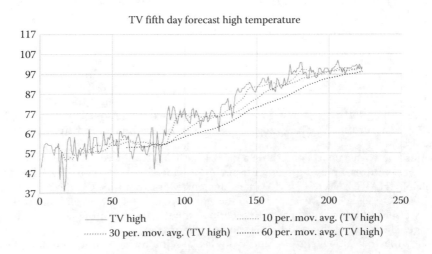

FIGURE 3.19

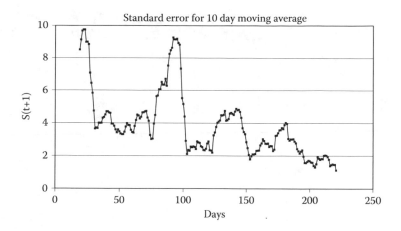

FIGURE 3.20

3.17.1 Standard Error

The standard error for the moving average function is defined by:

$$S(t+1) = \left\{ \Sigma\left[(A_{t+1-I} - F_{t+1-I})^2 / n \right] \right\}^{1/2} \tag{3.6}$$

This function has the same form as a population standard deviation.

The standard error for the 10-day moving average of Figure 3.18 is plotted in Figure 3.20. The decreasing trend with the approach of summer indicates less volatility in temperature as the calendar progresses. As one might expect, this simply means that Texas is predictably hot in the summer—day after day.

3.18 Bar and Column Charts

Although not as widely used as scatter charts, bar and column charts have a number of applications in engineering and are rather straightforward to create in Excel. The data are simply highlighted and the appropriate bar or column chart is created using INSERT/CHARTS/Bar Chart or INSERT/CHARTS/Column Chart. Editing with choices of fonts, fill patterns, line widths, etc., is essentially the same as with any other chart, but the editing of gap widths and overlap between columns deserves some special mention. To perform this editing, in either 2-D or 3-D bar or column charts, the data series are first activated by double-clicking on the chart. The Format Data Series window will then appear as shown in Figure 3.21 for a 2-D chart or Figure 3.22 for a 3-D chart.

For a 2-D bar or column chart, gap width is the spacing between the bars representing each data point. Overlap indicates the spacing between the adjacent data points. Negative Overlap indicates a space between the columns. Figure 3.23 illustrates the results of changing both parameters for a simple data system.

Similarly, in a 3-D bar or column chart, the parameters of gap depth, gap width, and chart depth may be varied to change the appearance of the final chart presentation.

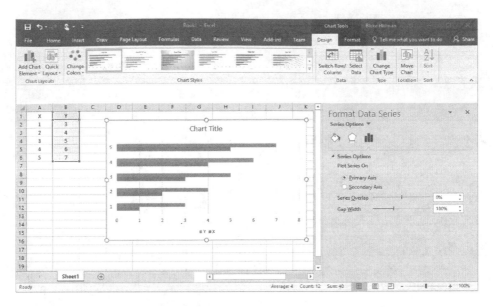

FIGURE 3.21

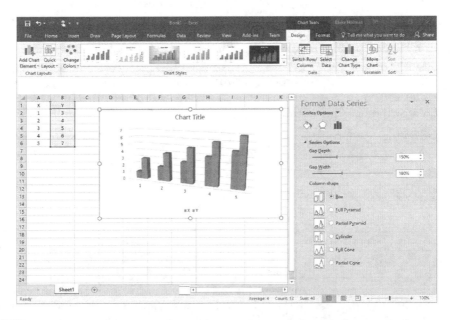

FIGURE 3.22

3.19 Chart Format and Cosmetics

Most charts prepared for engineering purposes will have a simple format involving minimal artistic or cosmetic effects. For visual presentations, color is certainly used to advantage. Excel offers the opportunity to adjust chart fills, fonts, colors, line size, and other effects.

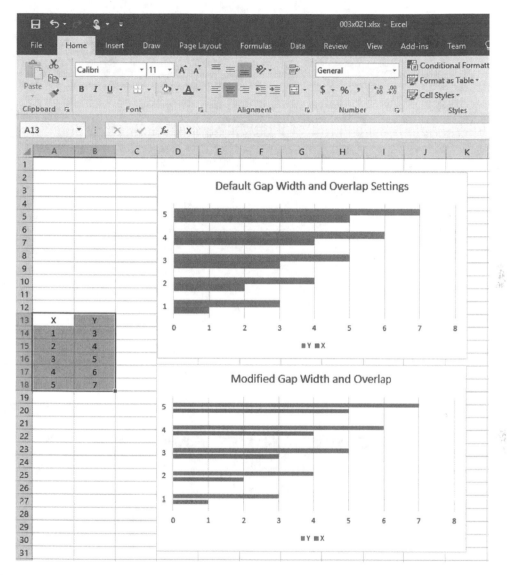

FIGURE 3.23

The purpose of this section is to illustrate the format windows that may be called up to make adjustments in various chart layouts and appearance. Figure 3.24 shows a very simple type 4 (Section 3.3) scatter chart. The main elements of the chart layouts that may be varied are Chart Area, Plot Area, Either Axis, Data Series, Title, and Gridlines. The format process is initiated by double-clicking one of these elements and thereby calling up the format window as shown in Figure 3.25. Note that the Chart Options drop-down menu has been selected to show all the chart elements that are editable through this window. For the Chart Area selection, there are Chart Options and there are Text Options that can be edited. For Chart Options, there are options for Fill & Line ( ), Effects ( ), and Size & Properties ( ). Clicking on each of these respective icons will present all the options from which to choose. The reader is encouraged to select each option area and become familiar with all available options.

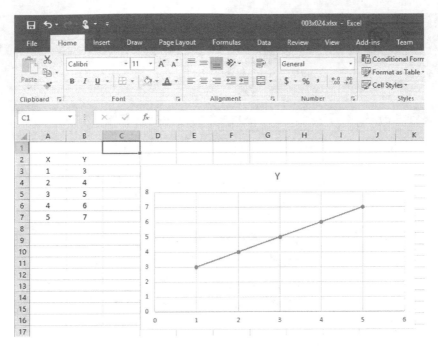

FIGURE 3.24

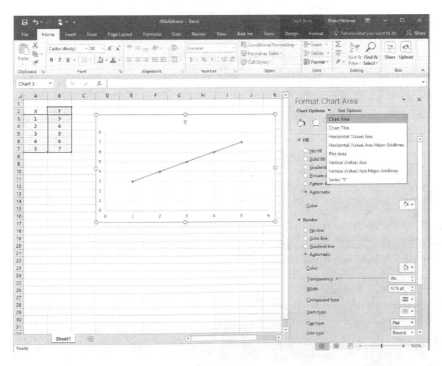

FIGURE 3.25

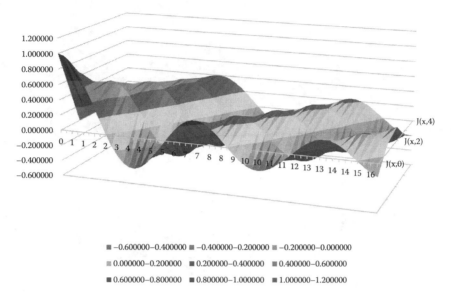

FIGURE 3.26

3.20 Surface Charts

Surface charts of the wire mesh or color variation type may be plotted by selecting a Surface Chart from the INSERT/CHARTS/Insert Surface or Radar Chart menu option. The adjustment of the chart depth is problematic, but may be accomplished using the following procedure:

1. Select (activate) the data.
2. On the Chart Wizard, choose Area Chart, 3-D effect.
3. Double click a chart element to bring up the Format window. On the Chart Options drop-down menu, select Side Wall and click on the Effects Icon. Scroll down and expand the 3-D Rotation options and increase the value of the Depth (% of base) from 100 to 520.
4. Change the chart type to Surface and resize the chart for better viewing.

Figure 3.26 shows the result of this procedure applied to the Bessel functions described in Section 3.13.

3.21 An Exercise in 3-D Visualization

This example gives the reader an opportunity to exercise his or her space visualization capabilities. Consider the set of 3-D views of the object shown in Figure 3.13.

Figure 3.27 displays the object in Figure 3.13e at several different elevation and rotation parameters. Before going further, the reader may want to examine each of these views and try to visualize their relative positions.

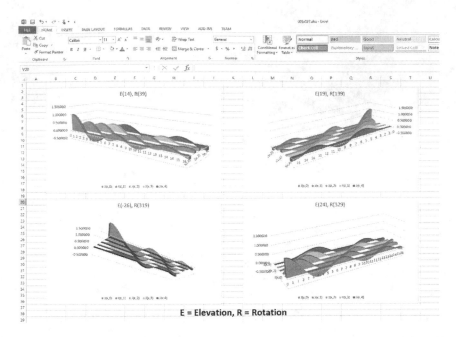

FIGURE 3.27

When considering the rotation of 3-D charts, note that certain angles represent certain views. Note that a rotation angle of 0° or 360° represents a view head-on or straight into the page. An elevation angle of 0° represents the same viewing position. An elevation angle of +90° represents a view straight down on the top of the object, whereas an elevation angle of –90° represents a view straight into the bottom of the object. The display of the views presented in Figure 3.27 notes the designations of their corresponding rotation and elevation angles.

Visualizing the different object positions of Figure 3.27 without the elevation and rotation information is not an easy task and represents some difficulty for most readers. This example illustrates once again the incredible display capabilities of Excel and the ease with which they can be accomplished.

Problems

3.1 The following data are collected in a certain experiment:

x	y
24,461	71.9
28,257	90.3
49,912	126.9
63,900	149.1
70,557	162
79,356	169
95,091	204
102,095	214
107,346	199.4
108,480	202.6

Plot the data as types 1, 2, and 4 scatter charts using linear, semi-log, and log–log coordinates. Based on these plots, obtain a suitable correlation for the data. Include the correlation equation and value of R^2 in each plot.

3.2 The following additional data are collected for the experiment of Problem 3.1. Add to the original data and obtain a new correlation for the complete set of data.

x	y
18,000	61
201,000	352
65,230	175
98,750	182

3.3 Plot the following data in a suitable scatter chart and obtain a trend line that best fits the data. Include the trend line, correlation equation, and value of R^2 in the chart.

x	y
0.2	0.1
0.5	0.3
1.2	1.5
2.4	5.9
3.1	8.9
4.6	21.2
5.1	24.9
6.9	42.6

3.4 Interchange the columns in Problems 3.1 through 3.3 and replot the data using y as the abscissa. Subsequently, obtain new correlation and trend line equations.

3.5 Two sets of variables are measured as functions of x and are tabulated as follows:

x	0	2	4	6	8	10	12
y_1	0.8	0.7	0.6	0.5	0.4	0.3	0.1
y_2	1	2	3	4	5	6	7

Plot the data on a type 4 scatter graph using (1) the same scale for y_1 and y_2, (2) an expanded scale for y_1, and (3) an expanded and inverted scale for y_1.

3.6 Plot the data of Problem 3.5 as (1) a column chart and (2) a bar chart.

3.7 Plot the data of Problem 3.5 as a surface chart with (1) variable surface color and (2) as a wire mesh chart.

3.8 Construct a table of the function:

$$y = e^{-0.1x} x \sin(nx)$$

for n = 1, 2, and 3 and $0 < x < \pi$. Select the increments in x as appropriate. Plot the function as (1) an area chart with 3-D, (2) a surface chart with variable colors, and (3) a 3-D wire mesh chart.

3.9   The following data are collected in an experiment:

x	y
4,000	27
5,000	42
6,000	38
7,000	50
8,000	45
9,000	50
10,000	49
20,000	92
30,000	120
40,000	115

Plot these data as types 1, 2, and 4 scatter graphs on linear coordinates. What do you conclude? Select the most appropriate of these plots and obtain linear, exponential, second-order polynomial, and power correlations of the data. Display the trend line, the correlation equation, and value of R^2 for each correlation. Depending on the results of these correlations, replot the data on semi-log or log–log coordinates to improve the data display.

3.10 Reconstruct the data of Figure 3.15. Plot the results as (1) an area chart with 3-D, (2) a surface chart with variable color, and (3) a 3-D wire mesh chart.

3.11 For colorful results, add the following fill effects to any of the charts obtained in the previous problems:

a. Fill the chart area with a colorful gradient of your choice by double clicking on the chart, selecting Fill from the Format Chart Area popup window, and then selecting the background pattern and colors that you like.

b. Fill the plot area with a colorful gradient of your choice by double clicking on the plot area of the chart, selecting Fill from the Format Chart Area popup window, and then selecting the background pattern and colors that you like.

3.12 A certain common investment stock has the following price history:

Period	Price
1	21
2	38
3	44
4	44
5	35
6	45
7	48
8	76
9	54
10	52
11	63

Plot the stock price as a function of period. Subsequently, construct the moving averages for the stock price having intervals ranging from two to four periods. Also, plot the standard error for each of the moving averages. Comment on the results.

3.13 The following results are calculated from a known analytical relationship:

x	y
1	6
2	16
3	35
4	58
5	85
6	122

Choose an appropriate scatter graph for plotting y as a function of x. Then, replot using y as a function of 1/x. Select the coordinate systems appropriate to the tabular values.

3.14 Plot the stock price data of Problem 3.12 as a column chart. Repeat for different gap widths and overlaps.

3.15 The following data are expected to follow a quadratic relationship. Investigate this expectation using an appropriate scatter chart and second-degree polynomial trend line fit.

x	y
0.1	3.667
1	3.724
10	4.223
100	7.247
1000	17.02

A quadratic function will plot as a straight line on linear coordinates when the ratio $(y - y_1)/(x - x_1)$ is plotted against x. Taking the second data set (1, 3.724) for the x_1 and y_1 coordinates, make such a plot and obtain a linear trend line fit to the data. How does this result compare with that obtained using the second-degree polynomial fit for the original data set?

4

Line Drawings, Embedded Objects, Equations, and Symbols in Excel

4.1 Introduction

In Chapter 3, we saw how it is possible to generate a variety of graphical displays in Excel, which may be employed for data presentation or calculation of results. The drawing capabilities of Excel offer further opportunities for displaying related schematic drawings or other information along with worksheet results and data manipulations. Although the drawing capabilities in Excel are not as extensive as in Computer Aided Design software or tools such as Microsoft Visio Professional, they are versatile and offer the convenience of creating drawings in other Microsoft Office documents. For those readers who use Microsoft PowerPoint, the drawing capabilities are even more useful.

Engineering schematics or drawings frequently involve the use of Greek or math symbols. The use of these symbols is much improved in recent versions of Excel. Examples and exercises in applying the various drawing capabilities will be given so that the reader can achieve some familiarity with these elements. The reader can then expand the use as his or her need dictates.

4.2 Constructing, Moving, and Inserting Straight Line Drawings

1. Open a new worksheet. Navigate to the Insert tab of the ribbon bar and focus attention on the Illustrations section.

2. Click Shapes and then click to select the Freeform line type under the Lines section of the Shapes menu. Figure 4.1 shows the Freeform line type icon.

3. Holding down the Shift key, click the crosshair at a point to start drawing with straight lines, quickly release-click, then move the crosshair to the next point and click again; repeat until the end of the drawing is reached and then double-click. If the end is at a closed figure, right click. Several line elements may be drawn separately to form the final drawing object, in which case multiple repetitions of this process must be performed. Line weight or style (including shading) may be adjusted by choosing the appropriate options from the Shape Styles section of the Format tab on the ribbon bar.

Lines

Freeform line type

FIGURE 4.1

The Excel worksheet grid may be used to guide the drawing process. Depending on the size of drawing needed, it may be advantageous to work with a reduced or compressed worksheet grid by adjusting the zoom level using the VIEW/ZOOM/ Zoom or VIEW/ZOOM/Zoom to Selection options. Reducing the column width with HOME/CELLS/Format menu choices may also help drawing precision. The row height may also be reduced to provide a finer drawing grid. Drawing pieces may be constructed separately and then dragged together to assemble an overall drawing object. Press F1 for Excel Help and enter the search term "Grouping" for more information in this regard. Precise movements of objects may be accomplished by selecting the object and pressing arrow keys for movement in the desired direction.

4. INSERT/SHAPES/Callouts may be used to add nomenclature, as can INSERT/ TEXT/Text Box, with or without arrows. Line borders of callouts may be removed by clicking the FORMAT/SHAPE STYLES/Shape Outline and then selecting No Outline. To avoid overlap of worksheet or graph gridlines, callouts or text boxes may be filled with white color.

5. Drawing objects with all annotations may be moved, copied, or inserted elsewhere by highlighting (selecting/dragging) the cells containing the object and then clicking HOME/CLIPBOARD/Copy or by using the keyboard shortcut CTRL+C. The target document is then opened, the desired location is clicked/selected (see upper-left cell for the location on an Excel worksheet). Click HOME/CLIPBOARD/ Paste and select the appropriate operation to insert the object into the target document. CTRL+V is a keyboard shortcut that can be used in the place of HOME/ CLIPBOARD/Paste.

4.2.1 Drawing Line Segments in Precise Angular Increments

To draw an unwavering straight-line segment (not a freeform object) in an angular increment of 15° (or precisely horizontal or vertical), click on the straight line icon (see Figure 4.1) from the INSERT/SHAPES/Lines menu. Then, while holding down the Shift key, click and hold the mouse button at the starting point and move the mouse to the end location to draw the line. When the Shift key is not depressed, the line may be drawn in any direction. This is very useful for drawing precise horizontal or vertical lines.

Example 4.1: Assortment of Drawing Shapes

Figure 4.2 shows a collection of different drawing shapes that may be constructed using the INSERT/SHAPES capabilities of Excel. The following remarks refer to the objects at the noted cell locations for items in which the construction is not obvious or already noted on the worksheet.

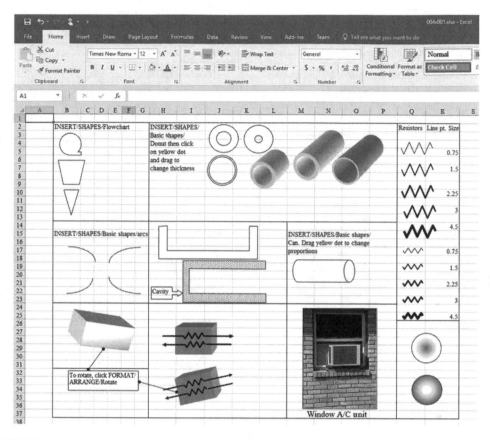

FIGURE 4.2

H2:P13—The donut shape is changed to a hollow cylinder using the 3-D effects tool, which allows variation in length. A gradient pattern fill is then added.

Q2:R25—A resistor shape is first drawn using the AutoShapes/Freeform lines tool (see discussion in Section 4.2, step 3) and then is copied several times using assorted line weights (as noted). It is also compressed by dragging.

B24:L37—A rectangle shape is drawn first. 3-D effects are added, and then a fill pattern with gradient effect is used.

Q26:R37—Two circles are drawn and then filled with gradient patterns from inner-to-outer and outer-to-inner.

H24:L37—A rectangle is drawn with 3-D, and light fill is applied. A resistor element is added along with arrows and straight lines. The elements are grouped (combined to form one object) and then rotated.

M24:P37—If needed, a digitized photo can be added. This digital photo is taken using a digital camera and copied to the worksheet. Editing of brightness, contrast, cropping, and image size may be accomplished before transfer to the worksheet. Editing can also be performed after transfer to the worksheet by right clicking the photo and selecting Format Picture from the popup menu.

Of course, the figure may also be presented without the presence of gridlines, and column and row headings.

Example 4.2: Construction, Assembly, and Labeling of a Line Drawing

An illustration of the mechanisms of an assembly of line drawings in Excel is shown in Figure 4.3:

1. In Figure 4.3a, a shell of a solar absorber is drawn with INSERT/SHAPES/ Lines/Free form as described previously.
2. Next, the inner boundary is drawn in Figure 4.3b and added to Figure 4.3a to produce the combination shown in Figure 4.3c. In practice, the two elements would not be moved around as shown here. The drawings are copied to show the steps as they progress.
3. A 5% pattern fill is added to the inner boundary as shown in Figure 4.3d. This represents the air inside the collector.
4. A long, thin rectangle is drawn in Figure 4.3e and filled with a 25% pattern fill. The dimensions of the rectangle are squeezed, adjusted, and moved to the top of the shell as shown in Figure 4.3f. This element represents the glass cover of the solar collector.
5. A 4.5-pt black line is added in Figure 4.3g to indicate the black absorbing surface at the bottom of the solar collector.
6. The elements of the assembled drawing are then grouped by (1) holding down Shift, (2) clicking on the elements in sequence, and (3) clicking FORMAT/ ARRANGE/GROUP/Group. The assembled object may now be moved or copied as a single object. In Figure 4.3i, the solar absorber has been rotated by

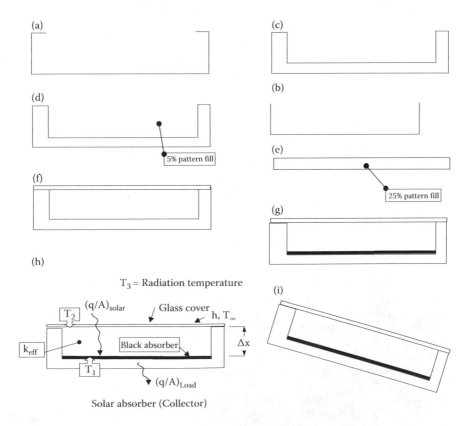

Solar absorber (Collector)

FIGURE 4.3

either clicking FORMAT/ARRANGE/Rotate or clicking the Rotate icon and dragging it to the desired angle.

7. The assembled drawing can then be copied to a Word document in which nomenclature is added with subscripts and appropriate symbols. Similarly, nomenclature can be added within Excel. The final diagram is shown in Figure 4.3h. The output shown here is without gridlines or column and row headings, although gridlines are very useful when constructing drawing elements.

Example 4.3: Practice Exercises with the INSERT and FORMAT Tabs

Gaining familiarity with the INSERT (Figure 4.4a) and FORMAT (Figure 4.4b) tabs shown in Figure 4.4 may be accomplished by carrying out the following exercises, which refer to the drawing objects shown in Figures 4.2 and 4.3. The solar absorber in Figure 4.3 is considered first.

For Figure 4.3a, click INSERT/SHAPES; then in the Lines section, click the Freeform icon (see Figure 4.1). Hold down the Shift key, move the crosshair to the desired starting point (a corner of a cell, for example), click quickly, and release the mouse button. While continuing to hold down the Shift key, move the crosshair to the next corner, click and release quickly, and continue until five segments (two top lips, two sides, and a bottom segment) are in place. Double click to complete the object.

For Figure 4.3b, perform the same operation as that shown in Figure 4.3a, except that only three line segments are required. Line up the drawing using a worksheet grid to obtain the dimensions in Figure 4.3b.

For Figure 4.3c, activate the drawing in Figure 4.3b and drag-move to match with Figure 4.3a, as shown in Figure 4.3c. Once the drawings are assembled, activate Figure 4.3b—click until boundary handles appear. Then, click the FORMAT/ SHAPE STYLES/Shape Fill menu to select a fill color and style. From this Shape Fill menu, you can select Gradients and then More Gradients to obtain a variety of Fill options. In this example, the Pattern fill option is used and the 5% fill option is selected. Selecting this will fill the object as in Figure 4.3d.

For Figure 4.3e, create a rectangle using the INSERT/SHAPES/Basic Shapes menu and selecting the appropriate rectangle. Click the mouse button to start the

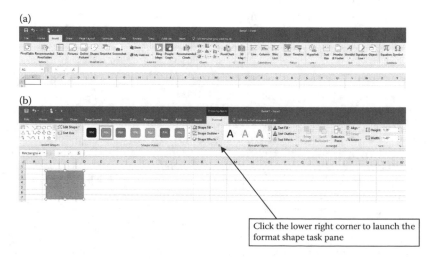

(a)

(b)

Click the lower right corner to launch the format shape task pane

FIGURE 4.4

rectangle and drag the mouse to draw the rectangle. Release the mouse button when the desired shape is attained. Do not worry if the rectangle is not the exact size or proportion needed. Next, use the FORMAT/SHAPE STYLES/ Shape Fill menu. Like above, select Gradients, More Gradients, and then Pattern fill to select the appropriate fill pattern. The fourth box on the top row of the pattern fill options will provide a 25% fill pattern. Click OK, and the rectangle will appear as in Figure 4.3e.

Drag the filled rectangle to a position on top of the drawing in Figure 4.3d. Then, drag, squeeze, or stretch the object to the dimensions shown. This is done by positioning the double-arrow pointer on the side handles of the filled rectangle by holding down the mouse button and dragging until the two figures line up. Nudge it into exact position using the arrow keys.

Next, draw a straight line at the inside bottom of the collector using INSERT/ SHAPES/Lines/Freeform as in Figure 4.3a and b. To create thickness for this line, use the Format Shape task pane to do so. To open the Format Shape task pane, click in the lower right corner of the FORMAT/SHAPE STYLES section (see Figure 4.4b). In the Format Shape task pane, under Line, adjust the Width by selecting the 4.5-pt line width.

Group the drawing elements created earlier by holding down the Shift key and clicking in sequence the four elements (a, b, e, and the 4.5-pt line). Click FORMAT/ARRANGE/GROUP/Group to connect these elements as one object. The composite object may now be moved as a single entity.

At this point, the composite figure may be copied to another Microsoft Office document by activating the object and then clicking HOME/CLIPBOARD/COPY. Open a Word or PowerPoint document and click HOME/CLIPBOARD/PASTE to paste the composite figure at the appropriate location within the document.

The nomenclature elements are added by using either text boxes or callout arrows (as for T_1 and T_2). These elements can be added directly from within Excel, or they can be added from within Word or PowerPoint. Arrows are simply another object type from within the INSERT/SHAPES choices. Subscripts or superscripts are formatting applied to text via the Font Settings menu and math symbols are inserted via the INSERT/SYMBOLS/Symbol menu option. See Section 2.3 for more detail in this regard.

After the nomenclature is added, group the nomenclature and the original composite drawing together via the FORMAT/ARRANGE/GROUP/Group menu option so that it appears as shown in Figure 4.3h.

Rotation of the object may be performed by activating the object and either clicking FORMAT/ARRANGE/Rotate or the Rotate icon (third from the left on the Drawing toolbar).

The following exercises refer to some of the drawing objects shown in Figure 4.2:

1. Create circles as shown in the figure at Q26:R37 using the INSERT/ ILLUSTRATIONS/Shapes and picking a circle. Holding down the Shift key, create a circle by moving the crosshair until the desired size is accomplished. Excel should automatically change to the Drawing Tools FORMAT tab on the ribbon bar. While the circle is still activated, click the FORMAT/SHAPE STYLES/Shape Fill menu item. Click the Gradient option that best matches the top circle. Repeat these steps to mimic the bottom circle.

2. Create donut shapes like those at J4 and K4. Use INSERT/ILLUSTRATIONS/ Shapes/Basic Shapes/Donut. Drag the yellow dot to achieve the desired thickness of the donut wall.

3. Create the hollow cylinder, as shown at N5, by first creating a donut. Then, click the FORMAT/SHAPE STYLES/Shape Effects/Bevel/3-D Options. Choose a

top and bottom bevel, and then increase the Depth setting to represent the desired length of the cylinder.

4. Create arcs like those at B17:G22 using INSERT/ILLUSTRATIONS/Shapes/ Basic Shapes/Arc. Then, repeat while holding down the Shift key. Note that circular arcs are created. Note the effect of dragging the yellow dot.

5. The resistor elements at Q2:R25 are created by (1) using INSERT/ ILLUSTRATIONS/Shapes/Lines/Freeform to construct a single resistor, (2) copying the resistor to other rows, (3) changing the line width of each copied resistor, (4) copying and compressing the resistor, and (5) subsequently copying the compressed resistor to other cells and modifying their respective line weights.

Performing these exercises enables a person to achieve a reasonable level of proficiency in drawing and formatting objects in Excel. Of course, further experimentation is encouraged.

4.3 Inserting Equation Templates and Symbols Using Excel and Word

Inserting symbols or equations in Excel, Word, or PowerPoint documents is greatly simplified by using the INSERT/SYMBOLS/Equation or INSERT/SYMBOLS/Symbol menu options. Word supports additional equation types beyond those supported in Excel that the reader can explore.

4.3.1 Symbol Insertion

For inserting symbols in Excel documents, use the following procedure:

1. Designate the location for a symbol insertion.
2. Click INSERT/SYMBOLS/Symbol to bring up the dialog box shown in Figure 4.5. From this dialog box, choose the font and then choose the symbol set to display the symbol options in the main window.
3. Select the desired symbol and press the Insert button.

Excel 2016 has 38 sets of symbols from which to choose using the dialog box mentioned in step 2. Additionally, Excel 2016 supports 27 special characters from the Special characters tab of the dialog box in step 2.

4.3.2 Equation Template Insertion

For inserting equations in Excel or Word documents, use the following procedure:

1. Designate the location for an equation insertion.
2. Click the down arrow on the INSERT/SYMBOLS/EQUATION menu to bring up the dialog box shown in Figure 4.6. From this dialog box, choose an equation template, which will insert an editable equation into the document. Custom equations

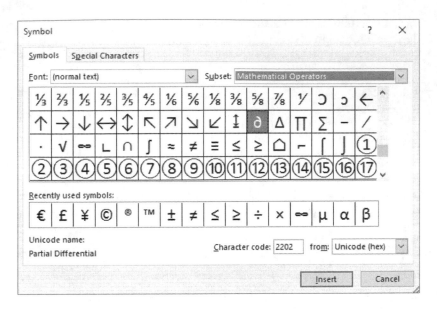

FIGURE 4.5

can be created using the Ink Equation menu option from INSERT/SYMBOLS/
EQUATION.

3. Edit the equation template to suit your needs.

4.4 Inserting Equations and Symbols in Excel Using Equation Editor

Inserting equations and symbols in Excel using the Equation Editor can be accomplished
by following this procedure:

1. Select a cell for the location of the equation.

2. Click INSERT/TEXT/OBJECT/Object, select Microsoft Equation 3.0 and click OK.

3. Before starting to construct an equation or symbol, set the size of the type by clicking SIZE/OTHER, and then enter the desired final point size. Click OK.

4. Construct the equation or symbols using features of Equation Editor.

5. When an equation or symbol has been constructed, double-click the object until it
   is highlighted or activated.

6. Click EDIT/CUT.

7. Click the cell or chart location for the insertion.

8. Click HOME/PASTE and select the appropriate paste operation. The object will
   appear at the designated cell or at the upper-left corner of the chart.

9. Activate the object and drag it to the final desired location. A box may enclose the
   object. Remove the line enclosing the box, if desired, by activating the object and
   clicking FORMAT/SHAPE STYLES/Shape Outline and clicking No outline.

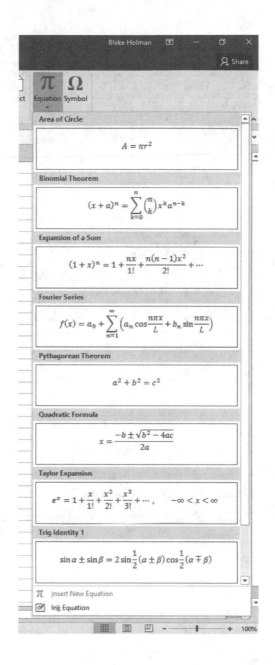

FIGURE 4.6

10. An empty box will appear at the cell chosen in step 1. Delete it.
11. To edit the object equation or symbol, double-click to activate it, and the Equation Editor tool bar will appear. Edit as appropriate.

Example 4.4: Graphics, Symbols, and Text Combinations

An example of the use of insertions and graphics composed and copied between Word and Excel is shown in Figure 4.7. In this figure, the assembly of items is printed as a

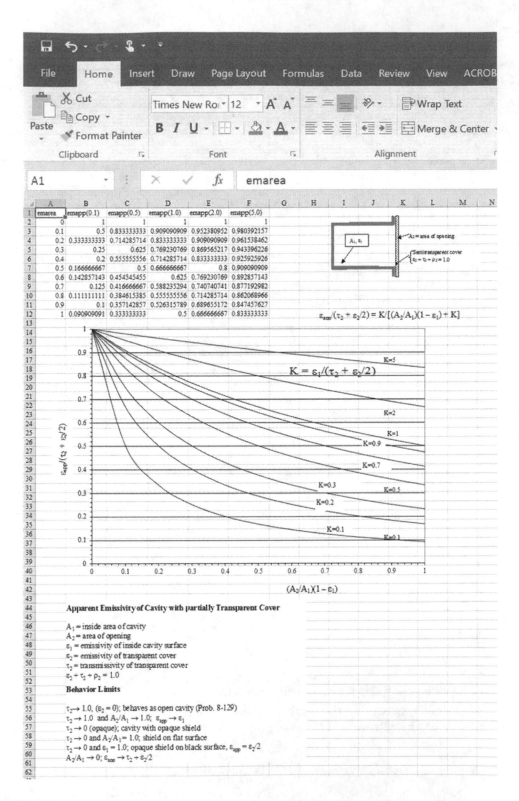

FIGURE 4.7

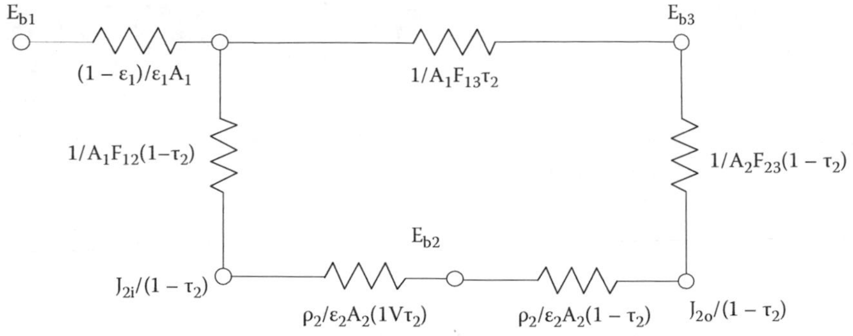

For $T_3 = 0$, $E_{b3} = 0$, and

$$q_1 = (E_{b1} - 0)/\Sigma R = \varepsilon_{app}A_2(E_{b1} - 0)$$

$$\varepsilon_{app} = 1/(A_2\Sigma R) = (\tau_2 + \varepsilon_2/2)K/[(A_2/A_1)(1 - \varepsilon_1) + K]$$

FIGURE 4.8

worksheet with column and row headings. Of course, the gridlines and row and column headings could have been left out to give a more appealing final presentation. Some elements of the assembly are shown in the following cell locations:

A1:F12—This is the Excel calculation of the equation at H12. The equation at H12 was created using the INSERT/SYMBOLS/Symbol operation.

H2:K10—This is the cavity drawing assembly. The cavity and semitransparent cover were drawn in Excel, grouped, and copied to Word. The labels using subscripts and Greek symbols were then added in Word, and the total assembly was copied back to the Excel worksheet. The labels, Greek symbols, and subscripts could equally have been created directly in Excel.

A14:L43—The graph was plotted from the Excel calculations using Chart Wizard with a type 3 x–y scatter graph chosen for the presentation (see Section 3.3). The x- and y-axis labels were composed within the Chart Wizard.

A44:H62—This descriptive write-up was composed in a text box within Excel, including symbol insertion and using subscripts.

Figure 4.8 shows a network schematic used to derive the equations for the illustration in Figure 4.7.

Example 4.5: Program with Embedded Text Documentation

Figure 4.9 illustrates a convenient approach to producing output in Excel. A program is presented at A2:D3 with variables and nomenclature at E2:F15. The documentation and description of the program can be written in a text box, including the use of Greek symbols, subscripts, and superscripts. This approach results in a compact presentation on a single page (if one can tolerate the small type size). A reader studying the program will find it much easier to follow than documentation occupying several pages.

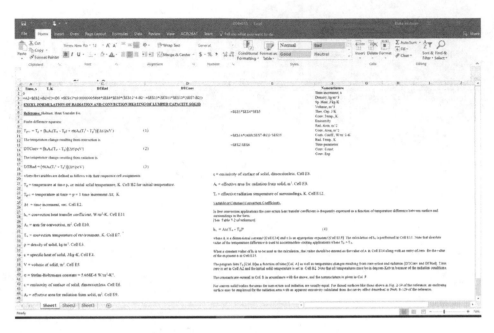

FIGURE 4.9

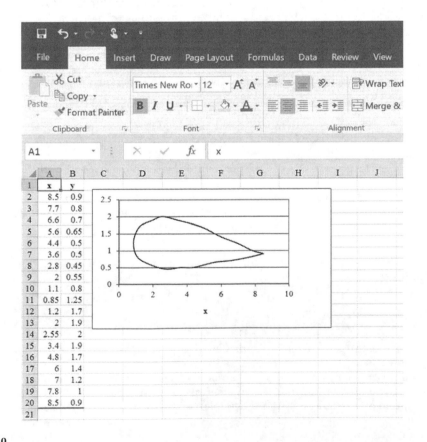

FIGURE 4.10

4.5 Construction of Line Drawings from Plotted Coordinates

A closed figure may be constructed in Excel by plotting a sequence of data points that returns to the initial coordinate set. An initial sketch on ordinary graph paper may form the basis for generating a smoothed line drawing when combined with the editing features of Excel graphs. We illustrate this technique with an airfoil shape. The airfoil shape has been hand-sketched on a sheet of graph paper graduated in increments of 0.1 in. The x- and y-coordinates of the shape are then tabulated in sequence, starting with the trailing edge of the airfoil, proceeding along the bottom surface of the section, and then around the top surface back to the trailing edge. The resulting data are shown in columns A and B of Figure 4.10, with the dimensions given in inches. A type 3 scatter chart is plotted and appears in Figure 4.10, without editing of the chart proportions. Note that the fitted computer curve for the tabulated data is not as smooth as one would expect for an airfoil section. This unevenness results from either a poor sketch (by the author) or inaccurate readings of the coordinates of the sketch.

The unevenness may be smoothed by first adjusting the chart proportions to the distorted form shown in Figure 4.11, which emphasizes the imperfections in the plot. Next,

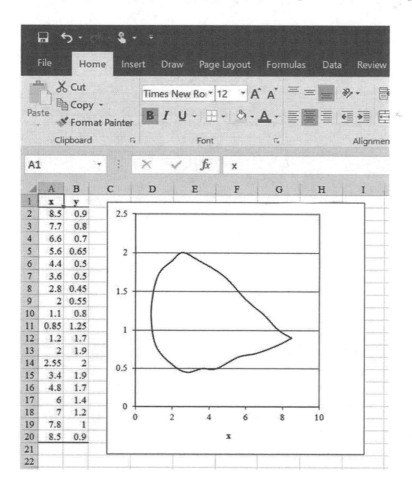

FIGURE 4.11

the data series is activated and the data points are double-clicked at points along the curve that appear to need adjustment. These points are then gently dragged to new positions to smooth out the curve. Two overcorrections are shown in Figure 4.12 to illustrate the action. The results of gentle smoothing are shown in Figure 4.13. Note that the results of the smoothing (or overcorrecting) also appear in the tabulated values for x and y.

Once the surface curve has been smoothed, the chart proportions are adjusted by dragging the side handles until the increments in x- and y-coordinates have the same chart measurements. The results are shown in Figure 4.14.

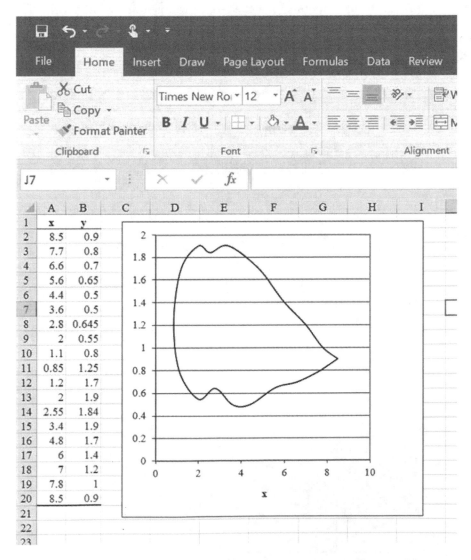

FIGURE 4.12

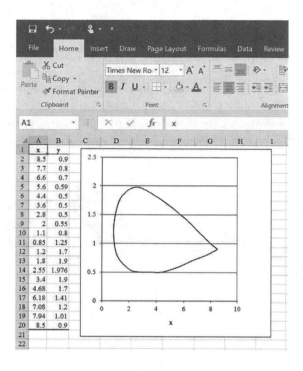

FIGURE 4.13

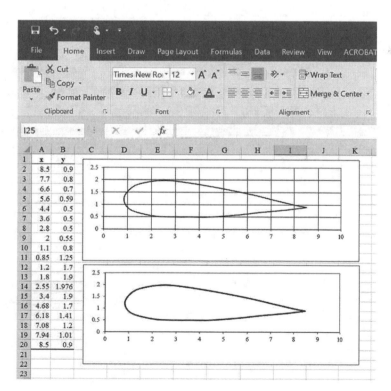

FIGURE 4.14

Problems

4.1 Plot the function $y = 3.25\ xe^{-0.1x}$ over the range $0 < x < 5$ using a type 3 scatter chart (Section 3.3). Create appropriate labels for the axes using the INSERT/SYMBOLS/Symbol menu Include the function on the appropriate axis label.

4.2 Construct the asymmetric circles shown. Then, fill the areas as indicated (or with a fill pattern of your choice), and engage the 3-D effects shown.

4.3 Create the text box shown with an inserted equation. Then, modify with the fill and shadow effects indicated. Note that activating the text box followed by FORMAT/ARRANGE/Rotate will allow you to orient the text box vertically.

4.4 Create Figure 4.8 by first creating the resistor elements as described in Section 4.3. Then, create the small circles. The resistor elements may be rotated by clicking FORMAT/ARRANGE/Rotate. Assemble the drawing in Excel and group with FORMAT/ARRANGE/GROUP/Group. Create and insert all the labels using text boxes without line borders. Also, construct the equations in text boxes below the diagram and position them appropriately with respect to the drawing.

4.5 If you have not already done so, work through the exercises in Section 4.3.

5

Solution of Equations

5.1 Introduction

In this chapter, we will examine the features of Excel that provide for solutions of single or simultaneous linear and nonlinear equations. Four methods will be described: (1) use of the Goal Seek feature, (2) use of the Solver feature, (3) iterative techniques, and (4) matrix inversion with the associated matrix operations. Examples will be given for each method and comments will be offered on the selection of the best method for a particular problem. Finally, a brief discussion will be presented on the creation of macros, along with an example.

5.2 Solutions to Nonlinear Equations Using Goal Seek

Nonlinear equations may be easily solved for real roots by using the Goal Seek feature, which is called by clicking DATA/FORECAST/What-If-Analysis/Goal Seek. First, the equation is written in the form

$$\partial(x) = 0$$

Keeping in mind that nonlinear equations may have multiple roots, including complex ones, it may be advantageous to plot the function to get an idea of the location of the possible roots. Goal Seek uses an iterative scheme to solve the equation, and an initial guess must be provided to start the computation. A graphical display may be useful in choosing the initial guess.

We consider two examples—a transcendental equation

$$\partial(x) = x \tan x - 2 = 0 \tag{5.1}$$

and a cubic polynomial

$$\partial(x) = 3x^3 - 2x^2 + x - 18 = 0 \tag{5.2}$$

The transcendental equation is plotted in Figure 5.1 using increments in x of 0.05 over the range $-2 < x < +2$. A visual survey of the graph indicates that there is a root at $x \approx 1.0$.

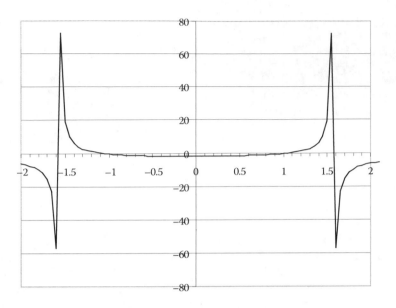

FIGURE 5.1

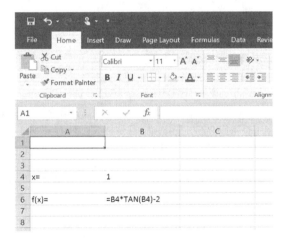

FIGURE 5.2

The worksheet in Figure 5.2 is set up with an initial guess for x inserted in cell B4 and the formula for $\partial(x)$ in cell B6. The guess of x=1.0 was chosen by consulting the plot in Figure 5.1. Next, DATA/FORECAST/What-If-Analysis/Goal Seek is clicked, which produces the window in Figure 5.4. We set cell B6=0 by changing (iterating) the values of x in cell B4. When OK is clicked, the window in Figure 5.5 appears along with the solution on the worksheet shown in Figure 5.3. Because of symmetry, there is also a root at x=−1.076845.

The same procedure is followed with the cubic polynomial. A graph of the function is shown in Figure 5.6, indicating a root at about x≈2 (it turns out that the root is exactly 2.0). The worksheet is set up as shown in Figure 5.7 with an initial guess taken as x=0. (We could have chosen x=2.0, but that would not be as interesting.) Again, DATA/FORECAST/What-If-Analysis/Goal Seek is called, and the solution is shown in Figure 5.8 having a value of x=1.999998558≈2.0.

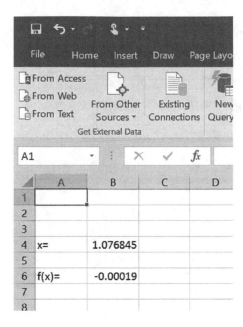

FIGURE 5.3

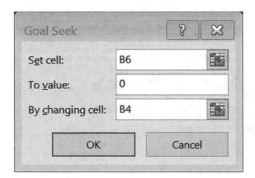

FIGURE 5.4

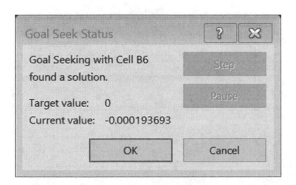

FIGURE 5.5

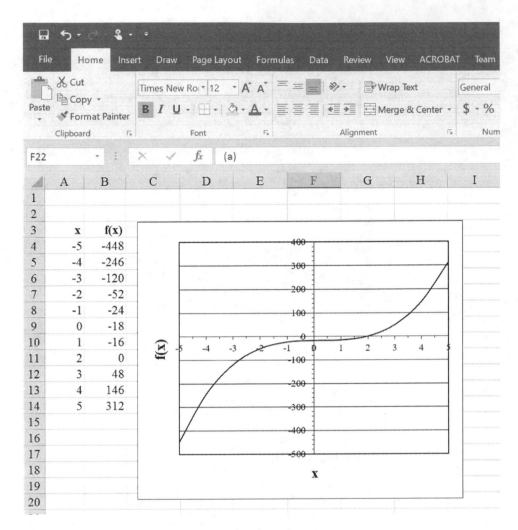

FIGURE 5.6

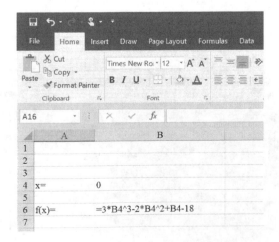

FIGURE 5.7

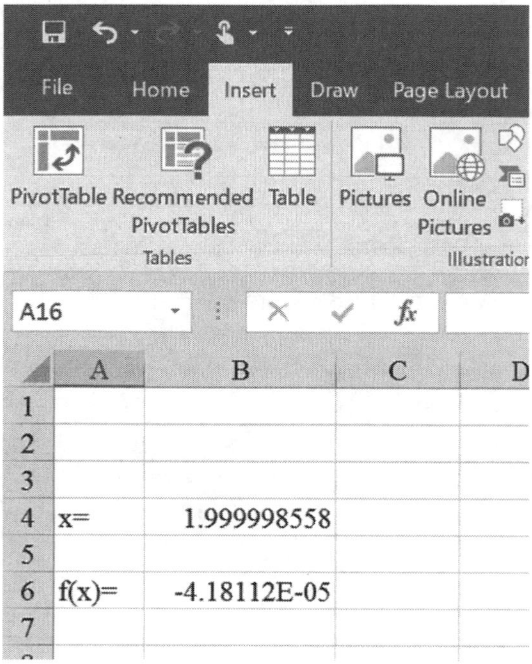

FIGURE 5.8

The graph for the cubic polynomial indicates that we should not expect any other real roots. Dividing the cubic polynomial by $(x-2)$ to extract the real root yields a quadratic function:

$$(3x^3 - 2x^2 + x - 18)/(x-2) = 3x^2 + 4x + 9$$

The roots of this quadratic function are complex and have the values $x = 2/3 \pm 1.5986i$.

5.3 Solutions to Nonlinear Equations Using Solver

Solver and Goal Seek offer alternate ways to solve nonlinear equations, although both employ iterative methods. A graph of the function is helpful in both instances because it indicates a reasonable value to use as the initial guess in the iterative process.

For Solver examples, we use the same nonlinear equations as used in the Goal Seek examples. In the top portion of Figure 5.9, we have the worksheet set up for the transcendental equation:

$$\partial(x) = x \tan x - 2 = 0$$

DATA/ANALYSIS/Solver is clicked, and the Solver Parameters window is displayed, targeting cell B4 to approach zero by changing the value of cell B3. An initial guess is listed as $x = 1.0$, and the solution is given in the right side of the top portion of Figure 5.9 as $x = 1.076874$ with a residual value of $\partial(x) = 4.29\text{E-7}$.

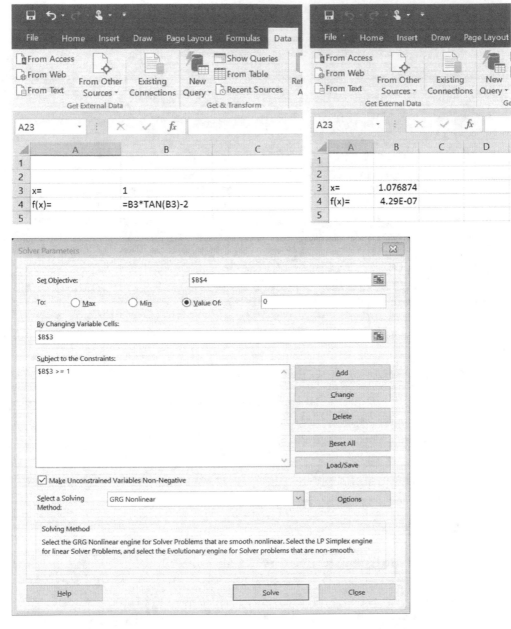

FIGURE 5.9

Figure 5.10 gives the worksheet for the cubic equation:

$$\partial(x) = 3x^3 - 2x^2 + x - 18 = 0$$

The target function is in cell B4, which is to approach zero by changing the x values in cell B3. An initial guess of x=1 is taken, and the result is given as x=2.00000001 with a residual value of $\partial(x) = 4.227E\text{-}6$.

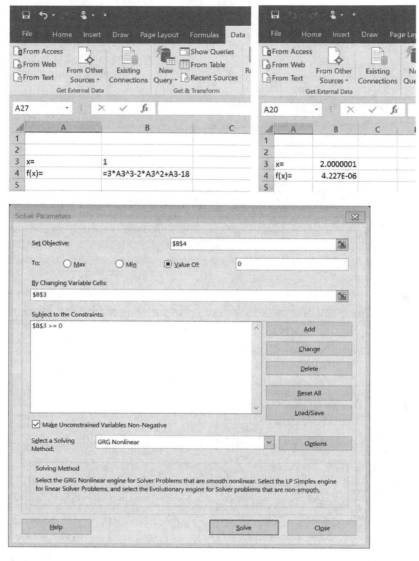

FIGURE 5.10

5.4 Iterative Solutions to Simultaneous Linear Equations

The following procedure may be used as an alternative to Gauss–Seidel or matrix solutions of simultaneous linear equations. It is particularly applicable to steady-state nodal equations in problems in which sparse coefficient matrices are involved.

1. Open a new Excel worksheet.
2. Click on FORMULAS/CALCULATION/Calculation Options and select Manual. This is necessary as you will be entering formulas that create cyclical references, which Excel does not like when set to auto-calculate.

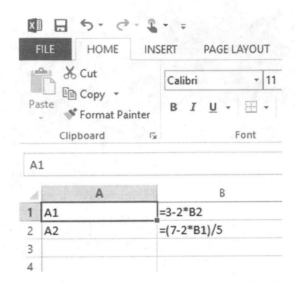

FIGURE 5.11

3. Toggle the worksheet into the mode of viewing formulas by clicking the CTRL+` key sequence.

4. Enter equations in the worksheet using the following format:

$$A_i = \partial(A'_j s)$$

For example, the equations

$$A_1 + 2A_2 = 3$$

$$2A_1 + 5A_2 = 7$$

would appear as shown in Figure 5.11.

5. Equations will now be in view on the worksheet. Check carefully to ensure there are no mistakes. Do not advance to the next step unless the equations are correct.

6. Press the CTRL+` key sequence to return to the numerical view on the worksheet. Solutions will now appear, in accordance with the number of iterations selected. The default value is 100 iterations with a deviation of 0.001. These values may be changed to obtain greater accuracy.

Example 5.1: Solution of Nine Nodal Equations

The following set of equations is obtained from a nodal analysis of a combined convection–conduction heat-transfer problem. Nine nodes were involved, so nine equations must be solved to obtain the temperature distribution in the solid. We have already written the equations in the format for an iterative solution.

$T_1 = (15.625 + 1.15T_2 + 1.15T_4)/5.425$
$T_2 = (1.15T_1 + 1.15T_3 + 31.25 + 2.3T_5)/10.85$
$T_3 = (1.15T_2 + 115 + 31.25 + 2.3T_6)/10.85$
$T_4 = (1.15T_1 + 2.3T_5 + 1.15T_7)/4.6$

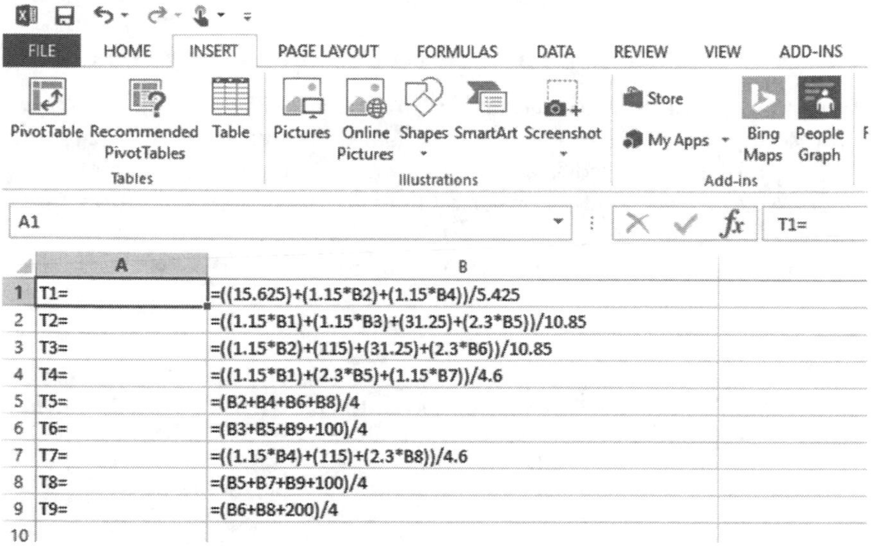

FIGURE 5.12

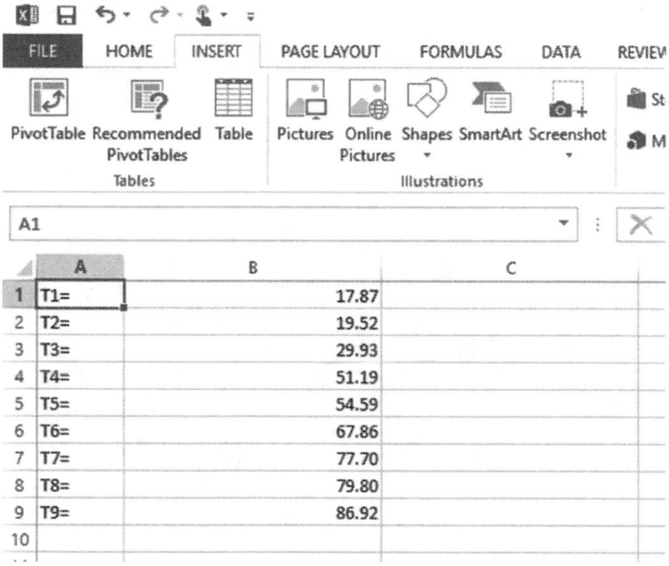

FIGURE 5.13

$$T_5 = (T_2 + T_4 + T_6 + T_8)/4$$
$$T_6 = (T_3 + T_5 + T_9 + 100)/4$$
$$T_7 = (1.15T_4 + 11_5 + 2.3T_8)/4.6$$
$$T_8 = (T_5 + T_7 + T_9 + 100)/4$$
$$T_9 = (T_6 + T_8 + 200)/4$$

The set of equations for a nine-node analysis of the problem is shown as columns A and B of the accompanying worksheet (Figure 5.12). A matrix of the nine equations would be sparse because each T does not connect with all of the other T's. The equations/solutions are displayed as a group by toggling CTRL+`. The solutions for the values of each T are shown in Figure 5.13.

5.5 Solutions of Simultaneous Linear Equations Using Matrix Inversion

We noted in solving the set of nine equations in Example 5.1 that a matrix solution to the problem would require entry of many more constants than the iterative solution, because that particular problem involved a very sparse matrix. For nonsparse matrices (or if one prefers, for sparse ones too), matrix inversion may be an attractive solution method. Excel provides an easy procedure for obtaining such solutions.

The set of linear equations may be written in the form

$$[A][X] = [C] \tag{5.3}$$

where [A] is the coefficient matrix, [X] is the set of unknown variables, and [C] is the constant matrix. The set of two equations

$$x_1 + 2x_2 = 5$$

$$3x_1 - 5x_2 = -7$$

will have

$$[A] = \begin{matrix} 1 & 2 \\ 3 & -5 \end{matrix}$$

$$[C] = \begin{matrix} 5 \\ -7 \end{matrix}$$

as the respective coefficient and constant matrices.

The solution is expressed as

$$[X] = [A]^{-1}[C] \tag{5.4}$$

where $[A]^{-1}$ is the inverse of the coefficient matrix, and $[A]^{-1}[C]$ is the product of the two indicated matrices.

Excel provides two worksheet functions to perform the matrix inversion and product operations. The function

MINVERSE(square array)

returns the inverse of the square array designated in the parentheses. Note that a square array is required for the MINVERSE() function. The function

MMULT(matrix 1, matrix 2)

returns the product of the two matrices in the parentheses. Unlike the MINVERSE() function, the matrix arguments to the MMULT() function need not be square arrays. Once the arguments are entered, pressing CTRL+Shift+Enter activates either of the matrix worksheet functions. If only a single value is returned instead of a matrix, you have failed to execute the CTRL+Shift+Enter input.

The procedure for obtaining matrix solutions to simultaneous linear equations is as follows:

1. Enter the coefficient matrix in a worksheet as a square array.
2. Determine the inverse of this array using the MINVERSE worksheet function.
3. Enter the constant matrix in the worksheet.
4. Multiply the results of step 2 by the array in step 3 to obtain the solution.

Example 5.2: Nine Nodal Equations Solution Using Matrix Inversion

The method may be illustrated by solving the nine-equation problem in Example 5.1 using matrix inversion. The worksheet is shown in Figure 5.14. First, the 9×9 array for the coefficient matrix is entered as [A] in the 81 cells B2:J10, whereas the constant matrix [C] is entered in the nine cells of B23:B31.

The inverse $[A]^{-1}$ is obtained by the following:

1. Activating cells B12:J20 to reserve space for $[A]^{-1}$.
2. Entering the formula=MINVERSE(B2:J10) in cell B12 at the top-left corner of the array.
3. Pressing CTRL+Shift+Enter to execute the inverse function.

The coefficients of $[A]^{-1}$ will appear in cells B12:J20 as shown.

Next, nine cells are activated at E23:E31 to reserve space for the solution matrix [T]. The formula

$$= MMULT(B12:J20, B23:B31)$$

is entered in the top cell E23, and upon pressing CTRL+Shift+Enter, the solution appears in cells E23:E31. It is understood that $T_1 = E23$, $T_2 = E24$, etc.

Finally, if one is interested, a check on the calculations may be made by executing the product [A][T] in the cells H23:H31. This is performed by activating these cells and entering the formula

$$= MMULT(B2:J10, E23:E31)$$

in the cell H23. The results in column H23:H31 should agree with those in column B23:B31. Note the agreement, although if you display enough precision on the result, cells H26 and H27 are not quite equal to zero because of the rounding errors in the calculation process.

Although we have shown both the matrix [A] and its inverse $[A]^{-1}$ in Figure 5.14, it is not necessary to calculate and display the coefficients of the inverse function $[A]^{-1}$ as an intermediate step in the solution process. Instead, one may go directly to the solution as shown in Figure 5.15. In this case, the inverse and matrix multiplication operations are performed by activating the nine cells E12:E20 to reserve space for the solution (Ts) and entering the following combination formula in cell E12:

$$= MMULT(MINVERSE(B2:J10), B12:B20)$$

where the cells B12:B20 represent the constant matrix [C].

	A	B	C	D	E	F	G	H	I	J	K
1											
2		-5.425	1.15	0	1.15	0	0	0	0	0	
3		1.15	-10.85	1.15	0	2.3	0	0	0	0	
4		0	1.15	-10.85	0	0	2.3	0	0	0	
5		1.15	0	0	-4.6	2.3	0	1.15	0	0	
6	[A]=	0	1	0	1	-4	1	0	1	0	
7		0	0	1	0	1	-4	0	0	1	
8		0	0	0	1.15	0	0	-4.6	2.3	0	
9		0	0	0	0	1	0	1	-4	1	
10		0	0	0	0	0	1	0	1	-4	
11											
12		-0.20650	-0.02952	-0.00556	-0.07504	-0.07642	-0.02638	-0.02720	-0.03882	-0.01630	
13		-0.02952	-0.10603	-0.01476	-0.03322	-0.09948	-0.03821	-0.01688	-0.03943	-0.01941	
14		-0.00556	-0.01476	-0.10047	-0.01147	-0.03821	-0.07310	-0.00709	-0.01941	-0.02313	
15		-0.07504	-0.03322	-0.01147	-0.32075	-0.26101	-0.08622	-0.11142	-0.14369	-0.05748	
16	$[A^{-1}]=$	-0.03322	-0.04325	-0.01661	-0.11348	-0.41197	-0.13051	-0.06247	-0.15688	-0.07185	
17		-0.01147	-0.01661	-0.03178	-0.03749	-0.13051	-0.32575	-0.02499	-0.07185	-0.09940	
18		-0.02720	-0.01688	-0.00709	-0.11142	-0.14369	-0.05748	-0.29355	-0.22219	-0.06992	
19		-0.01688	-0.01714	-0.00844	-0.06247	-0.15688	-0.07185	-0.09661	-0.37254	-0.11110	
20		-0.00709	-0.00844	-0.01006	-0.02499	-0.07185	-0.09940	-0.03040	-0.11110	-0.30262	
21											
22											
23		-15.625			17.86873			-15.625			
24		-31.25			19.51926			-31.250			
25		-146.25			29.93323			-146.250			
26		0			51.18759			0.000			
27	[C]=	0		[T]=	54.59206		[Check]=	0.000			
28		-100			67.86016			-100.000			
29		-115			77.69751			-115.000			
30		-100			79.80123			-100.000			
31		-200			86.91535			-200.000			
32											
33											

FIGURE 5.14

5.5.1 Error Messages

If any of the cells in [A] are left open, MINVERSE() will return the #VALUE! error value. It will also return this same error notice if the array does not have an equal number of rows and columns, i.e., if it is not a square array. If the determinant of [A] is zero, it is noninvertible and will return the #NUM! error value.

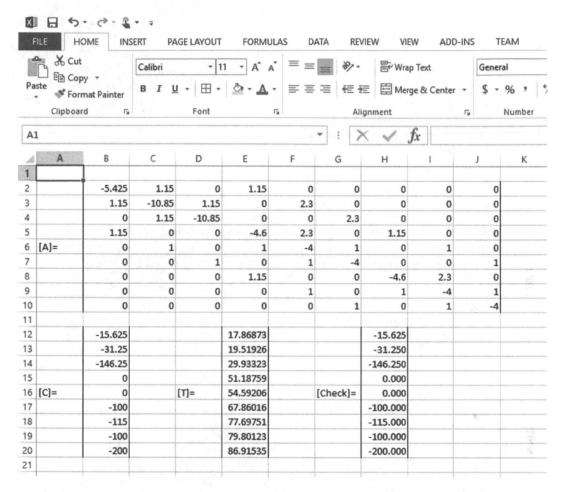

FIGURE 5.15

5.6 Solutions of Simultaneous Nonlinear Equations Using Solver

Excel Solver may be used to solve simultaneous linear or nonlinear equations. The equations are first written in the form:

$$f_1(x_1, x_2,\ldots, x_n) = 0$$

$$f_2(x_1, x_2,\ldots, x_n) = 0 \tag{5.5}$$

$$f_n = 0$$

In the equation set (Equation 5.5), there will be n equations in n unknowns. As noted, the equations may be linear or nonlinear. A new function $g(x_1, x_2,\ldots, x_n)$ is formed such that

$$g = f_1{}^2 + f_2{}^2 + \cdots + f_n{}^2 \tag{5.6}$$

The solution technique is to allow Excel Solver to iterate on values of x_1, x_2, etc., to cause the function g to approach zero. Because of the squares in the f functions, they too will approach zero and result in a solution for the set of equations. An alternate formulation would be to express g as a sum of the absolute values of the f function through $g = \Sigma ABS(f_i(x_i))$. For nonlinear equations, multiple sets of solutions may result (including complex solutions); hence, restrictions must be placed on the iterative process to match the physical problem represented by the equations. For example, a solution to a heat-transfer problem involving absolute temperatures would require that all the temperatures be positive. These restrictions must be inserted when formulating the problem.

The following is a suggested procedure for setting up the Excel worksheet to accomplish the solution:

1. In column A, type $x_1=$, $x_2=$, etc., for n rows.

2. In column B, adjacent to column A of step 1, insert initial estimates for values of x_1, x_2, etc., for n rows. These guesses should be made in accordance with the best estimate of the solution for the physical problem represented by the equations.

3. Skip a row, then in column A, type $f_1=$, $f_2=$, etc., for n rows, where the f's are in the form of a set of equations (Equation 5.5).

4. In column B, adjacent to column A of step 3, type functions ($=\cdots$) according to the equation set (Equation 5.5). Use cell locations from step 2 for designating the variables.

5. Skip a few rows, then in column A, type g=.

6. In column A, type g=.

7. In column B and the same row as step 6, type ($=\cdots$) function according to Equation 5.6. Use cell designations for the functions from step 4.

8. Click DATA/ANALYSIS/Solver. The target cell is that of the function in step 7. Set this target cell equal to zero by changing the cells in column B of step 2. Constraints are set according to physical problems. As noted, if absolute temperatures were the variables in a thermal problem, ≤ 0 might be set as the constraint.

9. Click Solve. A solution may or may not be obtained. If not, repeat Solve. Often, this will produce a solution. Or, alternate initial estimates for the variables in step 2 may be selected and the Solve procedure may be repeated. A solution usually results. If too high a "precision" is specified, Solver may state that a solution is not found when, in fact, one has been found but not to the precision selected. The person formulating the physical problem must make a judgment call in such cases. One should not ask for unreasonable limits of precision when the uncertainties of the physical problem do not justify them.

Example 5.3: Solution of a Set of Algebraic Equations

The worksheet and Solver windows are shown in Figure 5.16 for solutions of the following set of nonlinear equations:

$$x_1 + 3x_2 - 6x_3{}^2 = -47$$

$$7x_1 - 5x_2^3 + x_3 = -30$$

$$2x_1^2 + 4x_2 - 6x_3 = -8$$

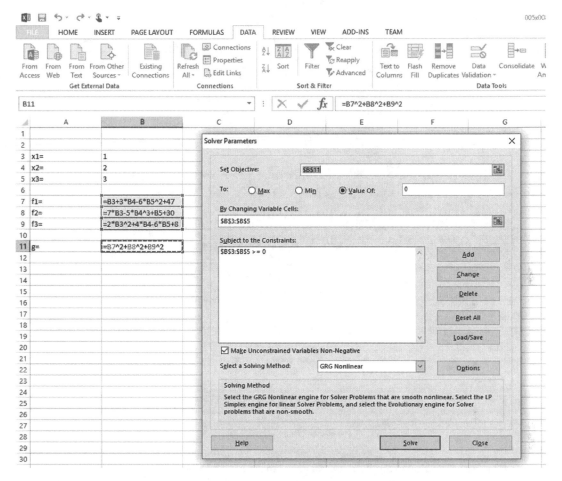

FIGURE 5.16

The resulting values of the f and g functions are also given. For this problem, the constraint of B3:B5≤0 (positive values) was selected. The exact solutions are $x_1=1$, $x_2=2$, and $x_3=3$. As an interesting exercise, try adjusting the desired precision of this example under the Options button of the Solver window and note the difference in solutions and values of the f and g functions.

Example 5.4: Radiation and Convection Heat Transfer between Two Plates

A system schematic of a heat transfer problem is shown in Figure 5.17. The temperature on the outside of the left plate is T_a, and the temperature on the outside of the right plate is T_b. The plates have different thickness and conductivities. Heat is conducted through each plate and is dissipated to fluid flowing between the plates. A fluid moves through the space between the plates at temperature T_f, and the inner surfaces of the plates exchange heat with each other by radiation energy, which is proportional to the absolute temperature to the fourth power. In addition, the plates lose heat by convection to the fluid. The convection coefficients are proportional to the temperature difference to the 0.25 power.

The energy balance on each inside surface is given in Equations 5.7 and 5.8 along with the temperature dependence of E_{b1}, E_{b2}, h_1, and h_2 in Equations 5.9 through 5.12. Our objective is to determine the temperatures of the inside surfaces of the walls, T_1 and T_2.

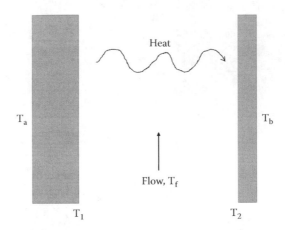

FIGURE 5.17

$$1000(T_a - T_1) + h_1(T_f - T_1) + (Eb_2 - Eb_1)/2.25 = 0 \tag{5.7}$$

$$5000(T_b - T_2) + h_2(T_f - T_2) + (Eb_1 - Eb_2)/2.25 = 0 \tag{5.8}$$

$$Eb_1 = 5.67E\text{-}8 \times T_1^4 \tag{5.9}$$

$$Eb_2 = 5.67E\text{-}8 \times T_2^4 \tag{5.10}$$

$$h_1 = 1.6 \times \left(ABS(T_1 - T_f)\right)^{0.25} \tag{5.11}$$

$$h_2 = 1.6 \times \left(ABS(T_2 - T_f)\right)^{0.25} \tag{5.12}$$

The purpose of this example is to illustrate the solution of nonlinear equations, so we ask the reader to accept the format of the equations as given. Detailed information on heat-transfer formulations is available in Reference 3.

The Excel worksheet is set up as shown in Figure 5.18, with the six variables located at B4:B9. The outside wall temperatures and fluid temperature are entered in column E with the values that are assigned for this particular case. Other values may be selected if the effects of different boundary conditions are to be examined. Equations 5.7 and 5.8 are already in the correct format [∂(…)=0], so they are entered in the worksheet at cells B11 and B12. The g function

$$g = f_1^2 + f_2^2$$

is entered in cell B14. This is the target cell that we want to iterate to zero by changing the values of T_1 and T_2 in cells B4 and B5.

Examining Equations 5.7 through 5.12, we see that the formulas in B6:B9 could have been incorporated in the formulas for f_1 and f_2. We choose to list them separately so that the calculated values of these quantities will become part of the solution presentation. The boundary temperatures in E4:E6 could also be inserted in the formulas; however, by using this type of display, they also become part of the solution presentation.

The formula displays are removed from the screen and the Solver window called by DATA/ANALYSIS/Solver. This window is shown in Figure 5.19. From the physical

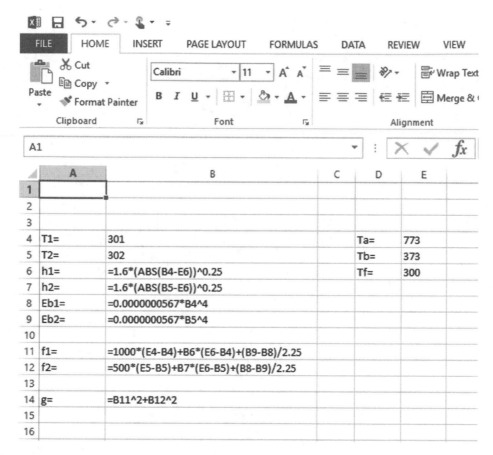

FIGURE 5.18

nature of the problem, it can be inferred that the minimum temperature in the two walls must always be greater than the fluid temperature $T_f = 300$ K; hence, the constraints are as shown in the Solver window. The initial guesses of 301 and 302 for T_1 and T_2 are shown at B4 and B5 of the formula window in Figure 5.18. All temperatures are expressed in absolute (degrees kelvin) because of the radiation terms.

After setting the target cell as B14 = 0, Solve is clicked, and the results are shown in Figure 5.19. Note the small value of g, i.e., $0.00005 \approx 0$.

Example 5.5: Solution of Simultaneous Linear Equations Using Solver

The procedure for solving a set of linear equations with Solver is the same as that for nonlinear equations. First, the equations must be written in the form $\partial(x_1,\ldots, x_n) = 0$. For this example, we choose the same set of equations that were solved by iteration in Example 5.1.

The listing of the T variables from B1 through B9 is shown in the worksheet in Figure 5.20. The equations for the ∂ functions are listed from B12 through B20 in this worksheet. Finally, the g function is listed as the sum of the squares of the ∂ functions in cell B22. The initial guesses for the T variables are all taken as zero in cells B1 through B9.

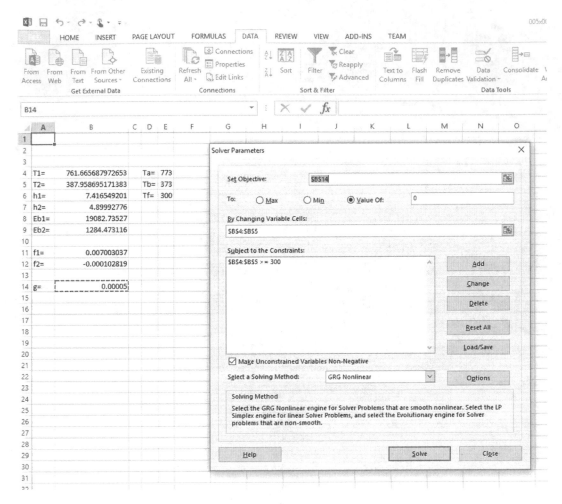

FIGURE 5.19

Solver is then called by clicking DATA/ANALYSIS/Solver. The Solver window appears as in Figure 5.20. The target cell that contains the g function is B22, which is set to zero. The cells to be changed for the iteration process are those of the variables in B1:B9. The constraint that the solutions be greater than or equal to zero is added. Solver is then clicked and a solution is found that appears as in Figure 5.21. All the ∂ functions and the g function have small values. The solutions for the T variables agree with the values obtained previously.

Although the use of Solver to obtain a solution to simultaneous linear equations is quite satisfactory, it is more cumbersome to use than the iterative technique described in Section 5.4.

5.7 Solver Results Dialog Box

The Solver Results dialog box offers additional options beyond keeping or rejecting a solution. When a solution is found, Solver also provides the user three types of reports:

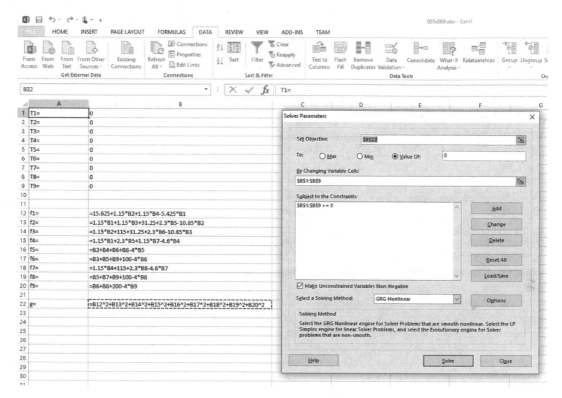

FIGURE 5.20

Answer, Sensitivity, and Limits reports. If Answer is checked, a summary answer report is generated and will appear as a separate Answer Report sheet in the workbook. Similarly, clicking Sensitivity and/or Limits will produce separate Sensitivity and/or Limits report sheets in the workbook. These reports are preformatted in a style suitable for presentation or transfer to other documents or presentations. The Answer report for Example 5.4 is shown in Figure 5.22. Further information regarding this feature is available via Excel Help. These reports are quite useful when presenting the results of optimization problems, as we will see in Chapter 8.

5.8 Comparison of Methods for Solution of Simultaneous Linear Equations

Three methods have been demonstrated for solving simultaneous linear equations: (1) an iterative technique particularly applicable to sets of equations with sparse coefficient matrices, described in Section 5.4, (2) an iterative technique using Excel Solver, described in Section 5.6, and (3) a solution using matrix inversion as described in Section 5.5. In physical problems, the constants in [C] often represent boundary conditions imposed on the problem, which may be varied to investigate their effects on the final solution. Their

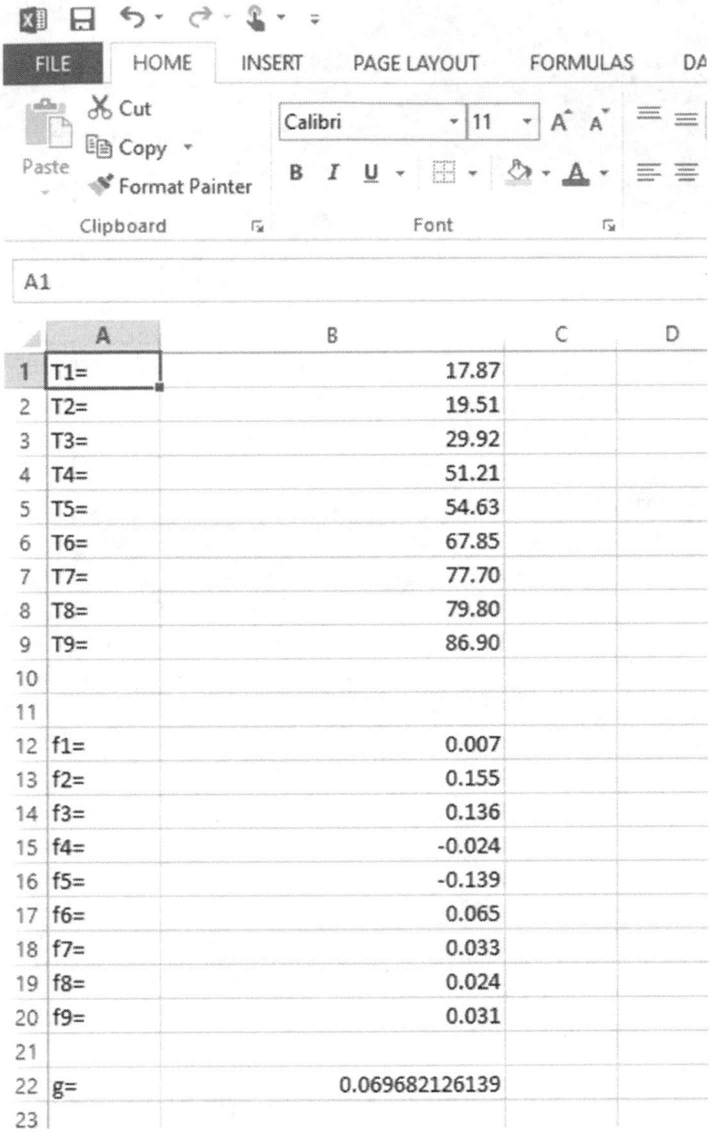

	A	B
1	T1=	17.87
2	T2=	19.51
3	T3=	29.92
4	T4=	51.21
5	T5=	54.63
6	T6=	67.85
7	T7=	77.70
8	T8=	79.80
9	T9=	86.90
10		
11		
12	f1=	0.007
13	f2=	0.155
14	f3=	0.136
15	f4=	-0.024
16	f5=	-0.139
17	f6=	0.065
18	f7=	0.033
19	f8=	0.024
20	f9=	0.031
21		
22	g=	0.069682126139
23		

FIGURE 5.21

display in the separate matrix offers the advantage of somewhat more convenient adjustments than when they are embedded in the equations for an iterative solution. Elements of the constant matrix may be expressed in terms of other cell addresses that may, in turn, be varied when examining different physical boundary conditions.

For an abundance of equations with sparse coefficient matrices, a large number of zeros must be entered. Over half the entries are zero in the nine-equation problem in Example 5.2. An error of just one entry will result in an incorrect solution. In these cases, the iterative technique is probably the most convenient solution method. Although not as visible as in the matrix method, the boundary conditions represented as constant terms in the equations may still be referred to variable cell locations and changed as needed in the physical problem.

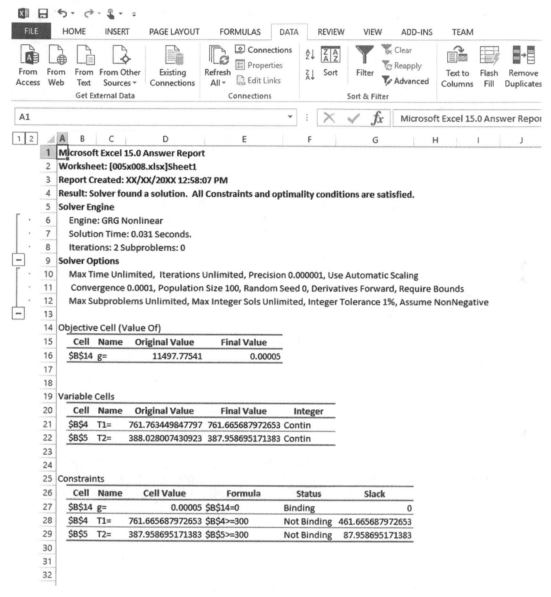

FIGURE 5.22

5.9 Copying Cell Equations for Repetitive Calculations

The drag-copy feature of Excel is very useful in applications in which repetitive calculations must be performed sequentially based on a previous result. One application is that of transient-heat-transfer analysis using finite-difference equations. Problems are usually formulated using the following nodal equation:

$$T_i^{p+1} = (\Delta\tau/C_i)\Sigma\left[(T_j^p - T_i^p)/R_{ij}\right] + T_i^p$$

where T^p represents temperatures at the beginning of a time increment, and T^{p+} represents nodal temperatures after a time increment $\Delta\tau$. A calculation of the transient temperature response of the object is performed by applying this equation sequentially to every node in the solid for as many time increments as desired. If the calculation is carried forward to a large number of time increments, the steady-state temperature distribution will be obtained. Different initial conditions may be examined easily by changing the temperatures that start the calculation at time zero.

Excel's reference address feature is leveraged to make formula creation easier by writing the cell (temperature node) formulas in reference address form and then copying them for as many time increments as needed. In this way, the p+1 node is always specified in terms of the p node preceding it in the worksheet. The procedure for this approach is as follows:

1. Select the number of time increments for the solution of the problem and set the number of rows required for setup equal to the number of time increments minus one. For example, 20 time increments will require 21 rows.

2. Enter the initial temperature conditions in the first row.

3. Enter the nodal equations in the worksheet in the required format, starting with row 2. Display the formulas in the body of the worksheet using the CTRL+` key sequence, which when used again will toggle back to displaying the worksheet values. Check carefully to see that the formulas are correct.

4. Copy the cell formulas down for the number of rows selected in step 1 by using the method outlined in section 2.9 (copying cells by dragging the Fill Handle).

5. Press the CTRL+` key sequence to display the temperature distribution on the worksheet.

6. The problem may be re-worked for other initial conditions by returning to step 2 and entering new values. The new solutions will appear immediately.

A simple numerical example of this method is given below. The reader unfamiliar with heat-transfer nomenclature should not be concerned with the formulas used, but instead should observe the behavior of the solution and the way the drag-copy operation is performed.

Example 5.6: Transient Temperature Distribution in a 1-D Solid

A 1-D slab is initially at a uniform temperature. The temperature of the left surface is suddenly changed to 100°C, while that of the right surface is suddenly changed to 200°C. Determine the temperature distribution at four positions in the solid using $(\Delta x)^2 / \alpha\Delta\tau = 2$ and a sufficient number of time increments to achieve steady state. Obtain solutions for initial temperatures of 0°C, 200°C, 300°C, and 1000°C.

Solution

When the transient parameter is selected as given, the nodal equations become (see Reference 3 for an explanation of this parameter):

$$T^{p+1}(m) = (1/2)\left[T^p(m+1) + T^p(m-1) \right]$$

where m+1 and m−1 refer to the temperatures to the right and left of node m, respectively.

Or, for row two of the worksheet, we have

$$A2 = (100 + B1)/2; \; B2 = (A1 + C1)/2; \; C2 = (B1 + D1)/2; \; D2 = (C1 + 200)/2$$

The equations are shown in Figure 5.23, and the numerical results are given in Figure 5.24 in the accompanying printouts for 40 time increments (41 rows), which are sufficient in all cases to achieve the steady-state values of 120°, 140°, 160°, and 180°C. Note that with the use of the drag-copy, the relative address feature expresses the temperatures in each row in terms of the temperatures in the previous row. Each row represents a time increment. The specific value of the time increment would depend on the parameters Δx and α, which are not defined in the problem statement.

5.10 Creating and Running Macros

A macro is a sequence of operations with keystrokes and mouse actions that may be recorded and stored for repeated use. The procedure for creating macros in Excel is most easily demonstrated with a specific example.

Example 5.7: Macro to Solve f(x)=0

Create a macro to obtain the roots to $f(x) = 0$, where $f(x)$ appears in cell B4 of the worksheet. Obtain the solution using Goal Seek, iterating the values of x contained in cell B3. Once the macro is created, different functions may be entered in B4 and a solution obtained with a single click action.

The procedure is as follows:

1. Set up a worksheet for the way it will be used—in this case, set up a worksheet for a Goal Seek solution of the function in cell B4, with variable x in cell B3.
2. This step consists of the following actions:
   a. Click VIEW/MACROS/Macros/Record New Macro. The Record Macro dialog box will appear as shown in Figure 5.25. Assign a name to the macro (Macro1, in this case). The first character of the macro name must be a letter, followed by your choice of other letters, numbers, or symbols. Spaces are not permitted, but an underscore may be used instead of a space between words.
   b. If desired, a shortcut key may be assigned for the macro at this time. The key must take the form of CTRL+letter. Numbers are not permitted.
   c. Specify the storage place of the macro. "This Workbook" is chosen for this example.
   d. Enter a description of the macro.
3. Click OK. To stop recording the macro, click VIEW/MACROS/Macros/Stop Recording.
   Note that by default, the recording of a macro will be based on absolute cell references. To change this behavior to relative cell references, click VIEW/MACROS/Macros/Use Relative References.
4. Execute the procedure to obtain the roots. In this case, the solution to B4=0 is performed as shown in the discussion of Goal Seek solutions. Obviously, the

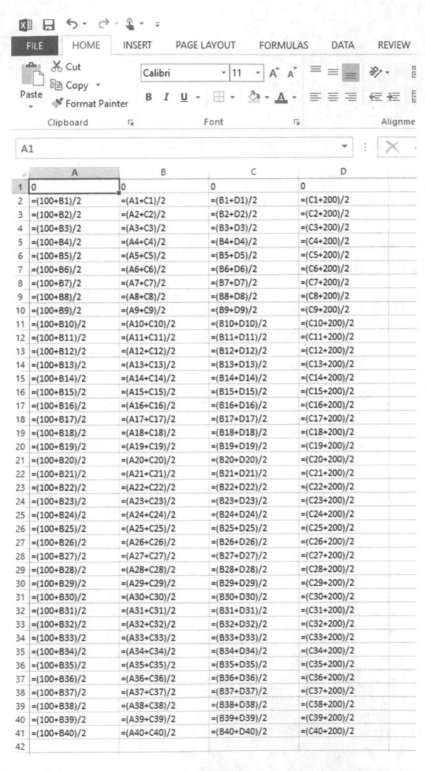

FIGURE 5.23

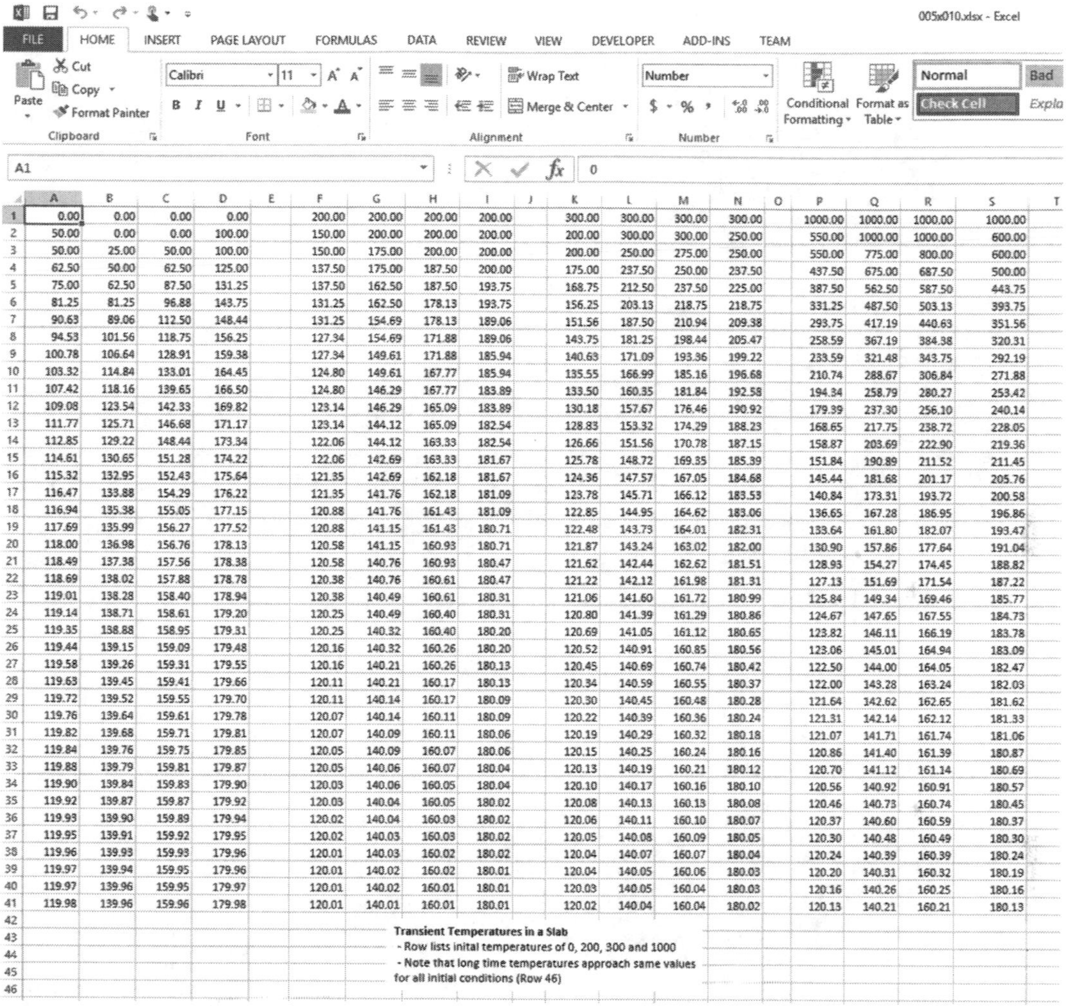

	A	B	C	D	E	F	G	H	I	J	K	L	M	N	O	P	Q	R	S	T
1	0.00	0.00	0.00	0.00		200.00	200.00	200.00	200.00		300.00	300.00	300.00	300.00		1000.00	1000.00	1000.00	1000.00	
2	50.00	0.00	0.00	100.00		150.00	200.00	200.00	200.00		200.00	300.00	300.00	250.00		550.00	1000.00	1000.00	600.00	
3	50.00	25.00	50.00	100.00		150.00	175.00	200.00	200.00		200.00	250.00	275.00	250.00		550.00	775.00	800.00	600.00	
4	62.50	50.00	62.50	125.00		137.50	175.00	187.50	200.00		175.00	237.50	250.00	237.50		437.50	675.00	687.50	500.00	
5	75.00	62.50	87.50	131.25		137.50	162.50	187.50	193.75		168.75	212.50	237.50	225.00		387.50	562.50	587.50	443.75	
6	81.25	81.25	96.88	143.75		131.25	162.50	178.13	193.75		156.25	203.13	218.75	218.75		331.25	487.50	503.13	393.75	
7	90.63	89.06	112.50	148.44		131.25	154.69	178.13	189.06		151.56	187.50	210.94	209.38		293.75	417.19	440.63	351.56	
8	94.53	101.56	118.75	156.25		127.34	154.69	171.88	189.06		143.75	181.25	198.44	205.47		258.59	367.19	384.38	320.31	
9	100.78	106.64	128.91	159.38		127.34	149.61	171.88	185.94		140.63	171.09	193.36	199.22		233.59	321.48	343.75	292.19	
10	103.32	114.84	133.01	164.45		124.80	149.61	167.77	185.94		135.55	166.99	185.16	196.68		210.74	288.67	306.84	271.88	
11	107.42	118.16	139.65	166.50		124.80	146.29	167.77	183.89		133.50	160.35	181.84	192.58		194.34	258.79	280.27	253.42	
12	109.08	123.54	142.33	169.82		123.14	146.29	165.09	183.89		130.18	157.67	176.46	190.92		179.39	237.30	256.10	240.14	
13	111.77	125.71	146.68	171.17		123.14	144.12	165.09	182.54		128.83	153.32	174.29	188.23		168.65	217.75	238.72	228.05	
14	112.85	129.22	148.44	173.34		122.06	144.12	163.33	182.54		126.66	151.56	170.78	187.15		158.87	203.69	222.90	219.36	
15	114.61	130.65	151.28	174.22		122.06	142.69	163.33	181.67		125.78	148.72	169.35	185.39		151.84	190.89	211.52	211.45	
16	115.32	132.95	152.43	175.64		121.35	142.69	162.18	181.67		124.36	147.57	167.05	184.68		145.44	181.68	201.17	205.76	
17	116.47	133.88	154.29	176.22		121.35	141.76	162.18	181.09		123.78	145.71	166.12	183.53		140.84	173.31	193.72	200.58	
18	116.94	135.38	155.05	177.15		120.88	141.76	161.43	181.09		122.85	144.95	164.62	183.06		136.65	167.28	186.95	196.86	
19	117.69	135.99	156.27	177.52		120.88	141.15	161.43	180.71		122.48	143.73	164.01	182.31		133.64	161.80	182.07	193.47	
20	118.00	136.98	156.76	178.13		120.58	141.15	160.93	180.71		121.87	143.24	163.02	182.00		130.90	157.86	177.64	191.04	
21	118.49	137.38	157.56	178.38		120.58	140.76	160.93	180.47		121.62	142.44	162.62	181.51		128.93	154.27	174.45	188.82	
22	118.69	138.02	157.88	178.78		120.38	140.76	160.61	180.47		121.22	142.12	161.98	181.31		127.13	151.69	171.54	187.22	
23	119.01	138.28	158.40	178.94		120.38	140.49	160.61	180.31		121.06	141.60	161.72	180.99		125.84	149.34	169.46	185.77	
24	119.14	138.71	158.61	179.20		120.25	140.49	160.40	180.31		120.80	141.39	161.29	180.86		124.67	147.65	167.55	184.73	
25	119.35	138.88	158.95	179.31		120.25	140.32	160.40	180.20		120.69	141.05	161.12	180.65		123.82	146.11	166.19	183.78	
26	119.44	139.15	159.09	179.48		120.16	140.32	160.26	180.20		120.52	140.91	160.85	180.56		123.06	145.01	164.94	183.09	
27	119.58	139.26	159.31	179.55		120.16	140.21	160.26	180.13		120.45	140.69	160.74	180.42		122.50	144.00	164.05	182.47	
28	119.63	139.45	159.41	179.66		120.11	140.21	160.17	180.13		120.34	140.59	160.55	180.37		122.00	143.28	163.24	182.03	
29	119.72	139.52	159.55	179.70		120.11	140.14	160.17	180.09		120.30	140.45	160.48	180.28		121.64	142.62	162.65	181.62	
30	119.76	139.64	159.61	179.78		120.07	140.14	160.11	180.09		120.22	140.39	160.36	180.24		121.31	142.14	162.12	181.33	
31	119.82	139.68	159.71	179.81		120.07	140.09	160.11	180.06		120.19	140.29	160.32	180.18		121.07	141.71	161.74	181.06	
32	119.84	139.76	159.75	179.85		120.05	140.09	160.07	180.06		120.15	140.25	160.24	180.16		120.86	141.40	161.39	180.87	
33	119.88	139.79	159.81	179.87		120.05	140.06	160.07	180.04		120.13	140.19	160.21	180.12		120.70	141.12	161.14	180.69	
34	119.90	139.84	159.83	179.90		120.03	140.06	160.05	180.04		120.10	140.17	160.16	180.10		120.56	140.92	160.91	180.57	
35	119.92	139.87	159.87	179.92		120.03	140.04	160.05	180.02		120.08	140.13	160.13	180.08		120.46	140.73	160.74	180.45	
36	119.93	139.90	159.89	179.94		120.02	140.04	160.03	180.02		120.06	140.11	160.10	180.07		120.37	140.60	160.59	180.37	
37	119.95	139.91	159.92	179.95		120.02	140.03	160.03	180.02		120.05	140.08	160.09	180.05		120.30	140.48	160.49	180.30	
38	119.96	139.93	159.93	179.96		120.01	140.03	160.02	180.02		120.04	140.07	160.07	180.04		120.24	140.39	160.39	180.24	
39	119.97	139.94	159.95	179.96		120.01	140.02	160.02	180.01		120.04	140.05	160.06	180.03		120.20	140.31	160.32	180.19	
40	119.97	139.96	159.95	179.97		120.01	140.02	160.01	180.01		120.03	140.05	160.04	180.03		120.16	140.26	160.25	180.16	
41	119.98	139.96	159.96	179.98		120.01	140.01	160.01	180.01		120.02	140.04	160.04	180.02		120.13	140.21	160.21	180.13	

Transient Temperatures in a Slab
- Row lists inital temperatures of 0, 200, 300 and 1000
- Note that long time temperatures approach same values for all initial conditions (Row 46)

FIGURE 5.24

solution to this example is very simple (B3=6). When the procedure execution is completed, click VIEW/MACROS/Macros/Stop Recording.

a. If the recording is done on an absolute cell basis (relative cell button not clicked), the solution will be obtained for B4=0 by changing the values of B3. The solution appears in B3 and the residual value of f(x) appears in B4.

b. If the macro is recorded on a relative cell basis (relative cell button activated), the function cell must be activated (clicked) before starting the macro. For example, if the function is entered in cell M9, that cell should be activated. The solution will then appear in the cell just above (or M8); therefore, that cell should be reserved for displaying the solution or initial guesses for the iterative Goal Seek calculation.

5. The macro is executed by pressing the shortcut key assigned (CTRL+letter) or by clicking VIEW/MACROS/Macros/View Macros, selecting Macro1, and then clicking Run. In addition, the macro may be attached to an object or button on the worksheet that will run the macro when clicked. Two examples are shown in Figure 5.26 for this macro.

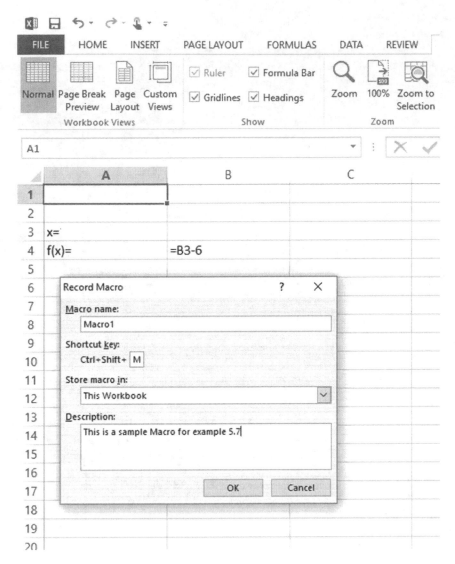

FIGURE 5.25

    a. A rectangle drawing object is created and the macro name Macro1 is typed inside. The macro is assigned by activating the rectangle, right clicking, followed by Assign Macro with Macro1 selected in the dialog box.

    b. A special button is created by clicking DEVELOPER/INSERT/Button. The Assign Macro box will appear. Make the assignment and then click OK. The button will remain activated; if not, activate by pressing CTRL while clicking on the periphery of the button. Type the title of the macro in the button using a desired style and font.

       Note that for this special button procedure, the user must enable the DEVELOPER toolbar. This is accomplished by selecting FILE/OPTIONS/ Customize Ribbon and then checking the Developer option on the right-hand side of the Customize Ribbon window.

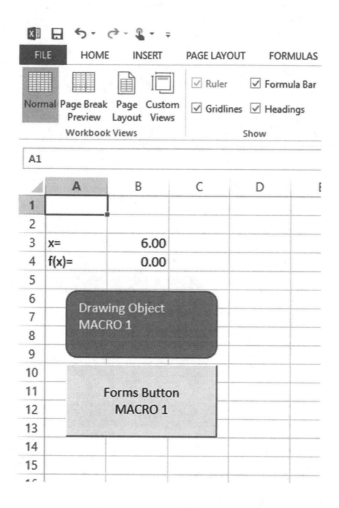

FIGURE 5.26

Problems

5.1 Solve the following equations using both Goal Seek and Solver:

a. $x = 0.09\left[1-(1+x)^{-n}\right]$ for $n = 5, 10, 15,$ and 20

b. $4x^3 - 3x^2 + 2x - 87 = 0$

c. $x\sin x - 1 = 0$

d. $xe^{-0.1x} = 1$

e. $x^3 - (0.647)^2\left[(1-x)^2(2+x)\right] = 0$

f. $3.587(1 + 0.04x^{2/3}) = 0.0668x$

g. $8.3(302 - x) = 5.102 \times 10^{-8}(x^4 - 278)$

h. $4.74(300 - x)^{1/4} + 5.102 \times 10^{-8}(500^4 - x^4) = 0$

5.2 Solve the following sets of linear equations using both the iterative technique and matrix inversion:

  a. $10x_2 + 10x_5 + 12.5 = 21.25x_1$

   $5x_1 + 5x_3 + 10x_6 + 12.5 = 21.25\ x_2$

   $5x_2 + 5x_4 + 10x_7 + 12.5 = 21.25x_3$

   $10x_3 + 12.5 = 11.25x_4$

   $10x_1 + 50x_6 + 12{,}000 = 100x_5$

   $25x_5 + 25x_7 + 10x_2 + 12{,}000 = 100x_6$

   $50x_6 + 20x_3 = 70x_7$

  b. $1100 + x_3 + x_4 = 4x_1$

   $600 + x_3 + x_4 = 4x_2$

   $900 + x_1 + x_2 = 4x_3$

   $800 + x_1 + x_2 = 4x_4$

  c. $75 + 2x_5 + x_2/4 + 16 = 3.3x_1$

   $x_1/4 + x_3/4 + 2x_6 + 16 = 3.3x_2$

   $x_2/4 + x_4/4 + 2x_7 + 16 = 3.3x_3$

   $x_3/4 + x_8 + 8 + 4 = 1.85x_4$

   $4x_1 + x_6/2 + 150 = 5x_5$

   $4x_2 + x_7/2 + x_5/2 = 5x_6$

   $4x_3 + x_6/2 + x_8/2 = 5x_7$

   $2x_4 + x_7/2 + 5 = 2.9x_8$

  d. $x_2/2 + 50 + 16x_3 = 17.75x_1$

   $x_1/2 + 16x_4 + 50 = 17.75x_2$

   $x_4 + 100 + 16x_5 + 16x_1 = 34x_3$

   $x_3 + 100 + 16x_6 + 16x_2 = 34x_4$

   $x_6/2 + 50 + 16x_3 = 17x_5$

   $x_5/2 + 50 + 16x_4 = 17x_6$

  e. $(57{,}000 - x_1)/4 + (460 - x_1)/90 + (x_2 - x_1)/19 = 0$

   $(x_1 - x_2)/19 + (460 - x_2)/31 + (x_3 - x_2)/64 = 0$

   $(460 - x_3)/8 + (x_2 - x_3)/64 = 0$

5.3 Solve the following sets of nonlinear equations using Solver:

a. $1300(T_2 - T_1) + 1.42[ABS(300 - T_1)]^{1/4} + 5.7 \times 10^{-8}(300^4 - T_1^4) = 0$

$T_3 + T_1 - 2T_2 = 0$

$500 + T_2 - 2T_3 = 0$

with the restriction that $300 < T < 500$

b. $1300[1 + 0.00025(T_2 + T_1)](T_2 - T_1) + 1.42[ABS(300 - T_1)]^{1/4} + 5.7 \times 10^{-8}(300^4 - T_1^4) = 0$

$[1 + 0.00025(T_3 + T_2)](T_3 - T_2) + [1 + 0.00025(T_1 + T_2)](T_1 - T_2) = 0$

$[1 + 0.00025(1000 + T_3)](1000 - T_3) + [1 + 0.00025(T_2 + T_3)](T_2 - T_3) = 0$

with the restriction that $300 < T < 1000$

c. $x_1^2 + \sin x - 2x_2 = 1.4674$

$x_1 x_2 + x_2^3 = 2.5708$

with the restriction that $x_1, x_2 > 0$

d. $x_1^2 + x_2^2 = 5$

$x_1 + 3x_2 = 7$

$x_2 + x_3^2 - x_1 = 5$

for all $x > 0$

e. $3.38 - p + [(101.3 - p)(310 - T)]/(1538 - T) = 0$

$\ln(p/2337) = 6789(1/293.15 - 1/T)$

p and T are positive values

5.4 The following set of equations describes the performance of a crossflow finned-tube heat exchanger:

$$e = 1 - \exp\{[\exp(-NCn) - 1]/Cn\}$$

$$n = N^{-0.22} C = 2100/C_{min}$$

$$DT_h = 0.67e \quad DT_h = 40,300/C_{min} \quad N = 2100/C_{min}$$

Determine the values of the six variables. All values must be positive.

5.5 The temperature ratio in a pin fin is described by the following equation:

$$Tr = [\cosh m(L-x) + (h/mk)\sinh m(L-x)]/[\cosh mL + (h/mk)\sinh mL]$$

where $m = (hP/kA)^{1/2}$; $P = \pi d$; $A = \pi d^2/4$

In a fin with d=0.01 m and L=0.1 m, the temperature ratios are measured at two x locations giving

$$Tr = 0.56 \text{ at } x = 0.04 \text{ m}$$

$$Tr = 0.365 \text{ at } x = 0.08 \text{ m}$$

Using Solver, determine the values of h and k.

5.6 The amplitude response for a seismic instrument is described by the following equation:

$$a = x^2 / \left[(1-x^2)^2 + (2Cx)^2 \right]^{1/2}$$

where $x = \omega/\omega_n$, $C = c/c_c$, $c_c = (4mk)^{1/2}$, and $\omega_n = (k/m)^{1/2}$

Three amplitude measurements are taken giving

$$a = 0.98 \text{ at } \omega = 75 \text{ Hz}$$

$$a = 2 \text{ at } \omega = 100 \text{ Hz}$$

$$a = 1.5 \text{ at } \omega = 166 \text{ Hz}$$

Using these data and Solver, determine values of m and k.

6

Other Operations

6.1 Introduction

In this chapter, we gather some operations that do not fall easily into the topics covered in other chapters. These items are as follows:

1. Numerical integration
2. Use of the logical IF function
3. Histograms
4. Normal error (probability) distributions
5. Calculation of uncertainty propagation in experiments
6. Multivariable linear and exponential regression analysis with LINEST and LOGEST worksheet functions
7. Examples and comparison of regression methods

Examples are given for each topic, and some coordination between the use of histograms, cumulative frequency distributions, and the normal error distributions is presented.

6.2 Numerical Evaluation of Integrals

Numerical evaluation of integrals may be performed in Excel by using either the trapezoidal rule or Simpson's rule. First, the area under the curve $y = \partial(x)$ is divided into increments of Δx. In the trapezoidal rule, the curve is replaced by a series of straight line segments, and the area of each element under the curve is calculated as a product of the mean height and the width Δx. Thus,

$$A_i = \Delta x(y_i + y_{i+1})/2$$

Taking the variables as ranging from 1 to n in indices, the total area under the curve will be

$$I = \int y dx = A = (1/2)(y_1 + 2y_2 + \cdots + 2y_{n-1} + y_n)\Delta x = [(y_1 + y_n)/2 + \Sigma y_i]\Delta x \qquad (6.1)$$

where the sum is carried out from i=2 to i=n−1.

If the increments in x are not uniform, the elemental areas are

$$A_i = y_m \Delta x_i = (y_i + y_{i+1})(x_{i+1} - x_i)/2 \tag{6.2}$$

The total area under the curve is then obtained by summing all the elemental areas.

Simpson's rule fits a series of parabolas to consecutive sets of three points on the curve such that the area may be calculated from

$$I = \int y dx = (y_1 + 4y_2 + 2y_3 + 4y_4 + 2y_5 + \cdots + 2y_{n-2} + 4y_{n-1} + y_n) \Delta x/3 \tag{6.3}$$

for uniform increments in x. If the area under the curve is divided into an even number of equally spaced values of Δx, the integral becomes

$$I = \left\{ y_i + y_n + \sum y_i \left[3 + (-1)^{i+1} \right] \right\} \times \Delta x/3 \tag{6.4}$$

where the summation is performed from $i=2$ to $i=n-1$. For an even number of increments in Δx, the number of data points will be odd.

The larger the number of increments in x, the better will be the approximation of the integral.

Example 6.1: Integration of the Sine Function

The worksheets shown in Figures 6.1 through 6.3 are set up to calculate the integral of sin(x) over the interval of $0 < x < \pi$. The exact value of the integral is equal to $-\cos(x)$

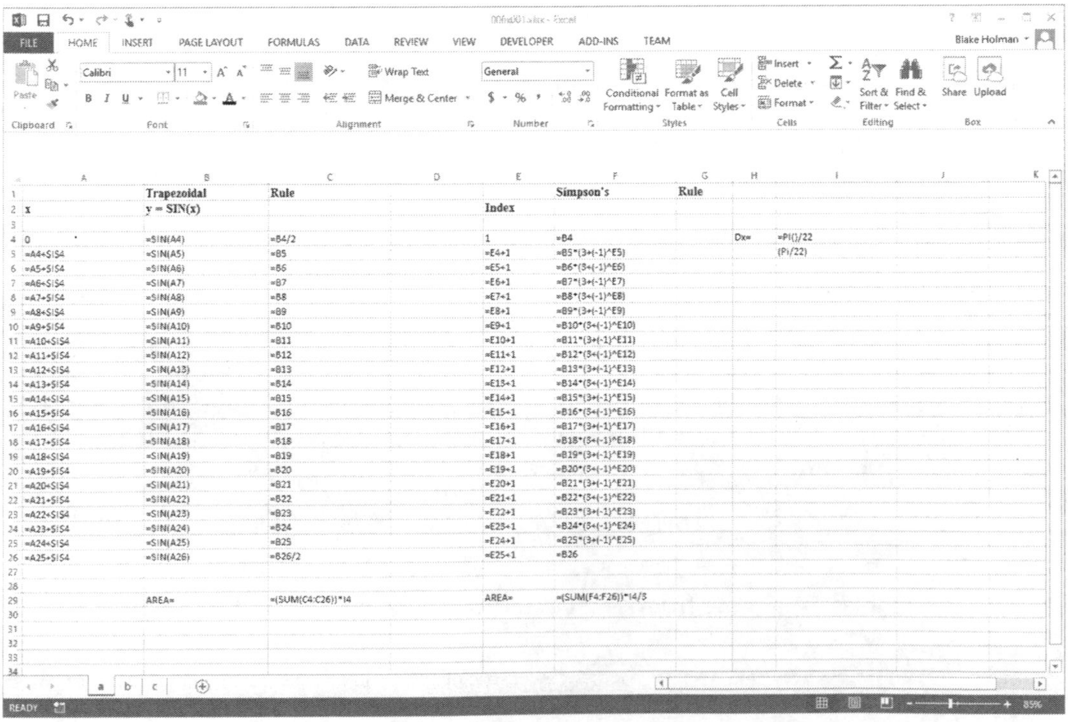

FIGURE 6.1

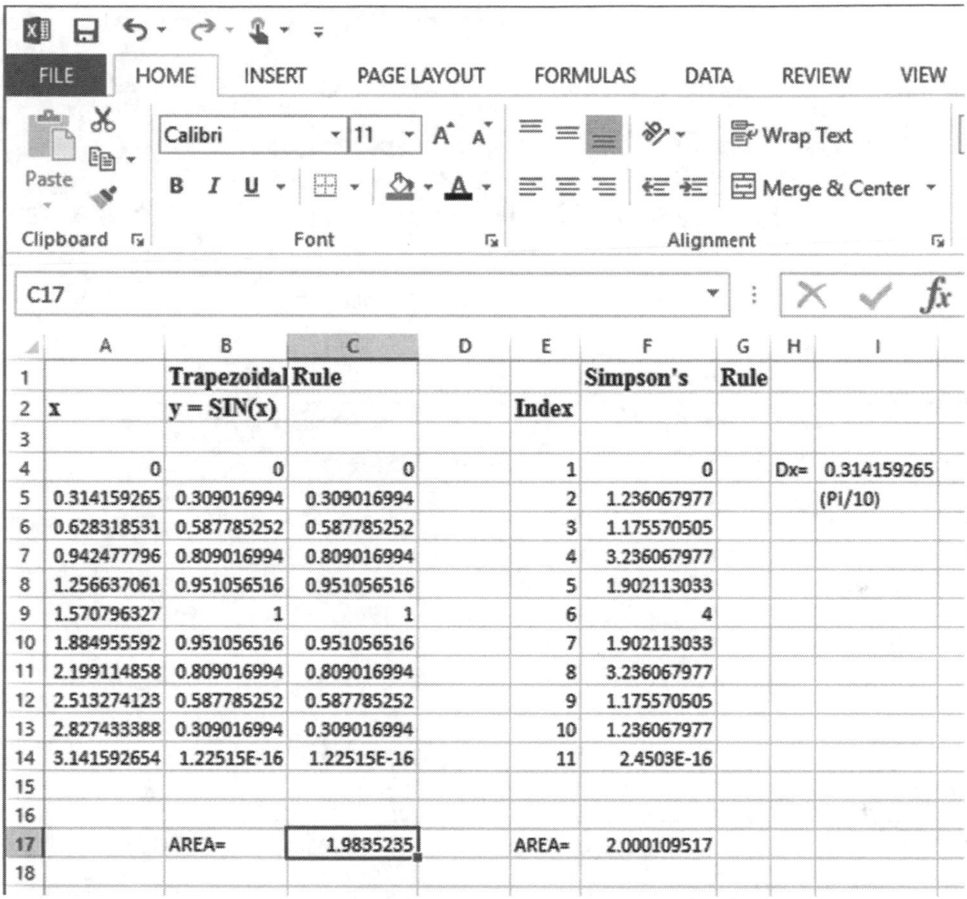

	A	B	C	D	E	F	G	H	I
1		**Trapezoidal**	**Rule**			**Simpson's**	**Rule**		
2	**x**	**y = SIN(x)**			**Index**				
3									
4	0	0	0		1	0		Dx=	0.314159265
5	0.314159265	0.309016994	0.309016994		2	1.236067977			(Pi/10)
6	0.628318531	0.587785252	0.587785252		3	1.175570505			
7	0.942477796	0.809016994	0.809016994		4	3.236067977			
8	1.256637061	0.951056516	0.951056516		5	1.902113033			
9	1.570796327	1	1		6	4			
10	1.884955592	0.951056516	0.951056516		7	1.902113033			
11	2.199114858	0.809016994	0.809016994		8	3.236067977			
12	2.513274123	0.587785252	0.587785252		9	1.175570505			
13	2.827433388	0.309016994	0.309016994		10	1.236067977			
14	3.141592654	1.22515E-16	1.22515E-16		11	2.4503E-16			
15									
16									
17		AREA=	1.9835235		AREA=	2.000109517			
18									

FIGURE 6.2

evaluated from 0 to π radians, which gives −(−1−1), or an exact value of 2.0. The worksheet is arranged to allow for different values of the increment Δx, which is assigned in cell I4. We will present the results for two Δx increments: π/10 and π/22. In both cases, we have an even number of increments, so Equation 6.4 is used for evaluation with Simpson's rule. In Figure 6.1, values of x are listed for the 26-increment case in column A, starting with zero and stepping up by increments specified in cell I4. Column B calculates the corresponding values of sin(x).

The first area (or integral) calculation is made using the trapezoidal rule of Equation 6.1. Half the values of y_1 and y_n are entered in cells C4 and C26, respectively, whereas the other y-values are copied in cells C5:C25 from cells B5:B25. Cell C29 then calculates the sum of the cells C4:C26 multiplied by the Δx increment to yield an area of 1.9966002 (0.016999% error) as shown in Figure 6.3.

For the calculation using Simpson's rule, the "i" index is created in column E and the arguments of the summation of Equation 6.4 are computed in cells F4:F26. Cell C29 then sums the entries of column F and multiplies by Δx/3 in accordance with Equation 6.4. The result for the calculated area is 2.000004631 (0.000231% error).

The same calculation is made for 10 increments in x, and the results are displayed in Figure 6.2. Use of fewer increments gives an area of 1.983523538 (−0.823825% error) for the trapezoidal rule and 2.000109517 (0.005476% error) when Simpson's rule is applied. In both cases, Simpson's rule is more accurate than the trapezoidal rule.

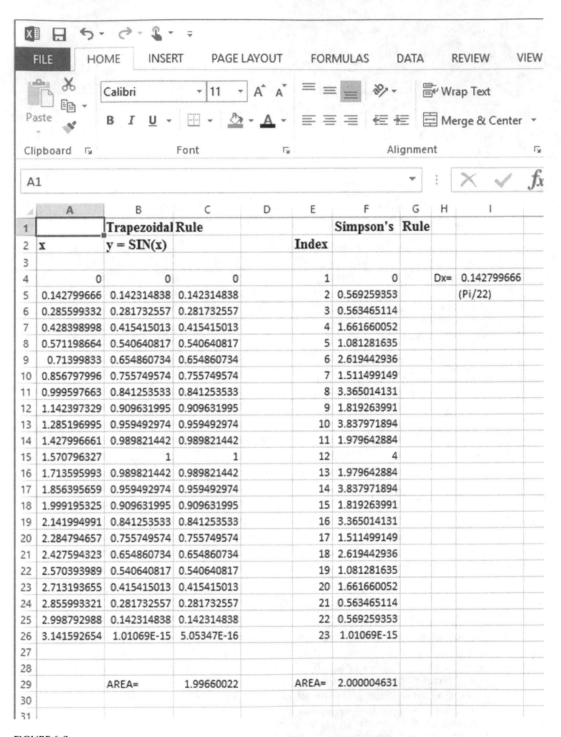

	A	B	C	D	E	F	G	H	I
1		Trapezoidal Rule				Simpson's Rule			
2	x	y = SIN(x)			Index				
3									
4	0	0	0		1	0		Dx=	0.142799666
5	0.142799666	0.142314838	0.142314838		2	0.569259353			(Pi/22)
6	0.285599332	0.281732557	0.281732557		3	0.563465114			
7	0.428398998	0.415415013	0.415415013		4	1.661660052			
8	0.571198664	0.540640817	0.540640817		5	1.081281635			
9	0.71399833	0.654860734	0.654860734		6	2.619442936			
10	0.856797996	0.755749574	0.755749574		7	1.511499149			
11	0.999597663	0.841253533	0.841253533		8	3.365014131			
12	1.142397329	0.909631995	0.909631995		9	1.819263991			
13	1.285196995	0.959492974	0.959492974		10	3.837971894			
14	1.427996661	0.989821442	0.989821442		11	1.979642884			
15	1.570796327	1	1		12	4			
16	1.713595993	0.989821442	0.989821442		13	1.979642884			
17	1.856395659	0.959492974	0.959492974		14	3.837971894			
18	1.999195325	0.909631995	0.909631995		15	1.819263991			
19	2.141994991	0.841253533	0.841253533		16	3.365014131			
20	2.284794657	0.755749574	0.755749574		17	1.511499149			
21	2.427594323	0.654860734	0.654860734		18	2.619442936			
22	2.570393989	0.540640817	0.540640817		19	1.081281635			
23	2.713193655	0.415415013	0.415415013		20	1.661660052			
24	2.855993321	0.281732557	0.281732557		21	0.563465114			
25	2.998792988	0.142314838	0.142314838		22	0.569259353			
26	3.141592654	1.01069E-15	5.05347E-16		23	1.01069E-15			
27									
28									
29		AREA=	1.99660022		AREA=	2.000004631			
30									
31									

FIGURE 6.3

Example 6.2: Numerical Integration of Experimental Data

We now examine a hypothetical set of experimental data shown as the variables x and y at the top left of the worksheet of Figure 6.4. The data are expected to follow a linear variation and hence are plotted on a linear x–y scatter graph as shown in Figure 6.4a through d. The graph in Figure 6.4a is a type 5 scatter graph with straight line segments joining the data points, the graph in Figure 6.4b is a type 3 scatter graph with a computer-smoothed curve joining the points, and the graph in Figure 6.4c is a type 1 scatter graph. The graph in Figure 6.4d is the same as that in Figure 6.4c but with the addition of a linear trend line fit for the data.

If a linear plot is expected from either physical reasoning or previous experience, then Figure 6.4d may be the preferred vehicle for presentation of the data. It is interesting to perform a numerical integration of the data over the range $0 < x < 5$ with increments $\Delta x = 0.5$, the same as used for the data increments.

The integration is performed similar to that of the sine function in Figure 6.1 using both the trapezoidal rule and Simpson's rule. In Figure 6.5, the formulas are displayed, whereas in Figure 6.6, the computed values of the integral in cell C19 for the trapezoidal integration and in cell F19 for Simpson's rule are shown. The computed values are 24.15 and 23.8666667, respectively. It is interesting to compare these numbers with those obtained by integrating over the trend line of Figure 6.4d.

The trend line is represented by

$$y = 1.9145x + 0.0318$$

and the area under the curve by

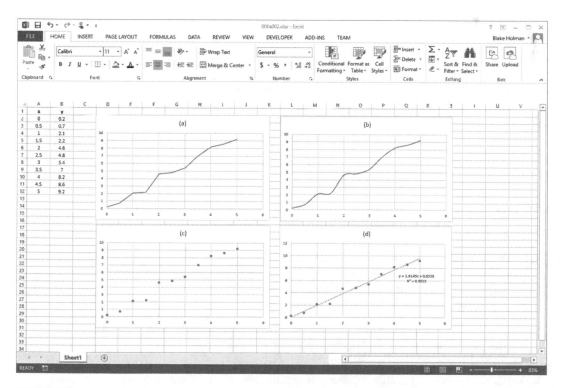

FIGURE 6.4

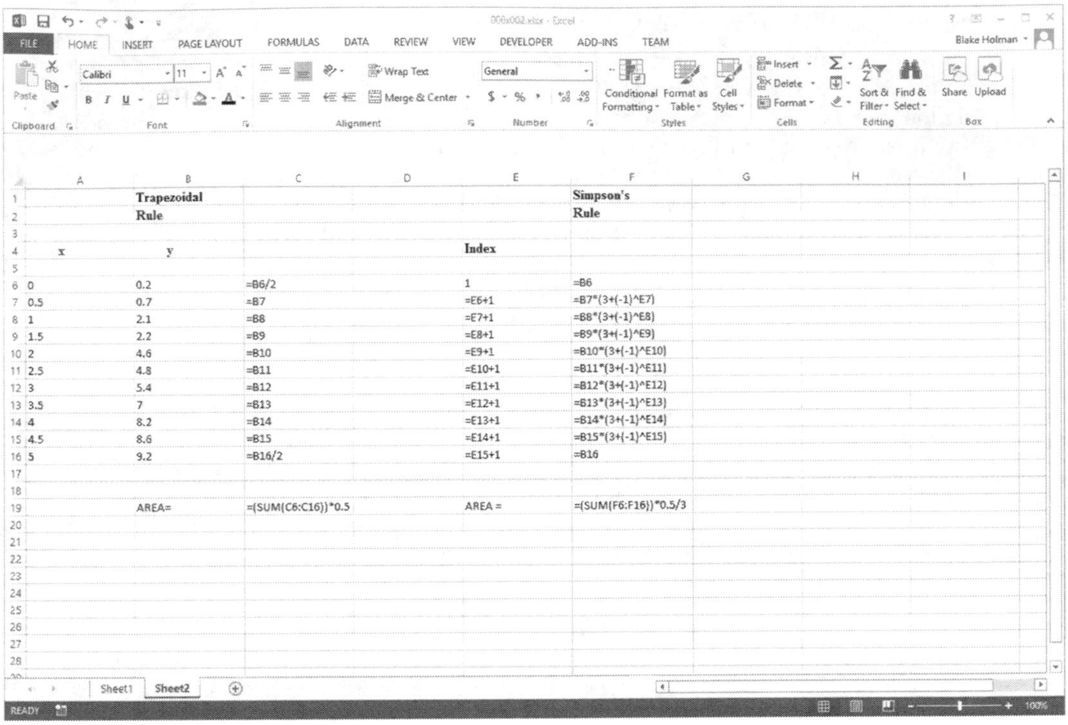

	A	B	C	D	E	F
1		Trapezoidal				Simpson's
2		Rule				Rule
3						
4	x	y			Index	
5						
6	0	0.2	=B6/2		1	=B6
7	0.5	0.7	=B7		=E6+1	=B7*(3+(-1)^E7)
8	1	2.1	=B8		=E7+1	=B8*(3+(-1)^E8)
9	1.5	2.2	=B9		=E8+1	=B9*(3+(-1)^E9)
10	2	4.6	=B10		=E9+1	=B10*(3+(-1)^E10)
11	2.5	4.8	=B11		=E10+1	=B11*(3+(-1)^E11)
12	3	5.4	=B12		=E11+1	=B12*(3+(-1)^E12)
13	3.5	7	=B13		=E12+1	=B13*(3+(-1)^E13)
14	4	8.2	=B14		=E13+1	=B14*(3+(-1)^E14)
15	4.5	8.6	=B15		=E14+1	=B15*(3+(-1)^E15)
16	5	9.2	=B16/2		=E15+1	=B16
17						
18						
19		AREA=	=(SUM(C6:C16))*0.5		AREA =	=(SUM(F6:F16))*0.5/3

FIGURE 6.5

$$\text{Area} = \int y\,dx = \left[1.9145x^2/2 + 0.0318x\right]_0^5 = 24.09025$$

This value compares favorably with the results obtained in the numerical integration shown earlier, and may be the best representation of the data because it is determined by the trend line fit. The smoothed curve in Figure 6.4b is probably unrealistic, and the scatter of the data is best taken into account by the least-squares trend line fit in Figure 6.4d.

6.3 Use of Logical IF Statement

The IF statement will place a value in the cell it occupies based on a true or false result of a logical test. The function takes the form:

IF (logical test, place value of ____ in the current cell if test is true; otherwise, place value of ____ in the current cell if test is false)

Example 6.3: Nested IF Statements and Embedded Documentation

Figure 6.7 shows the worksheet for a calculation that uses two nested IF functions (up to 64 are permitted, although not advised) to choose the proper calculation equation for flat-plate heat-transfer coefficients. The three equations are listed in mathematical

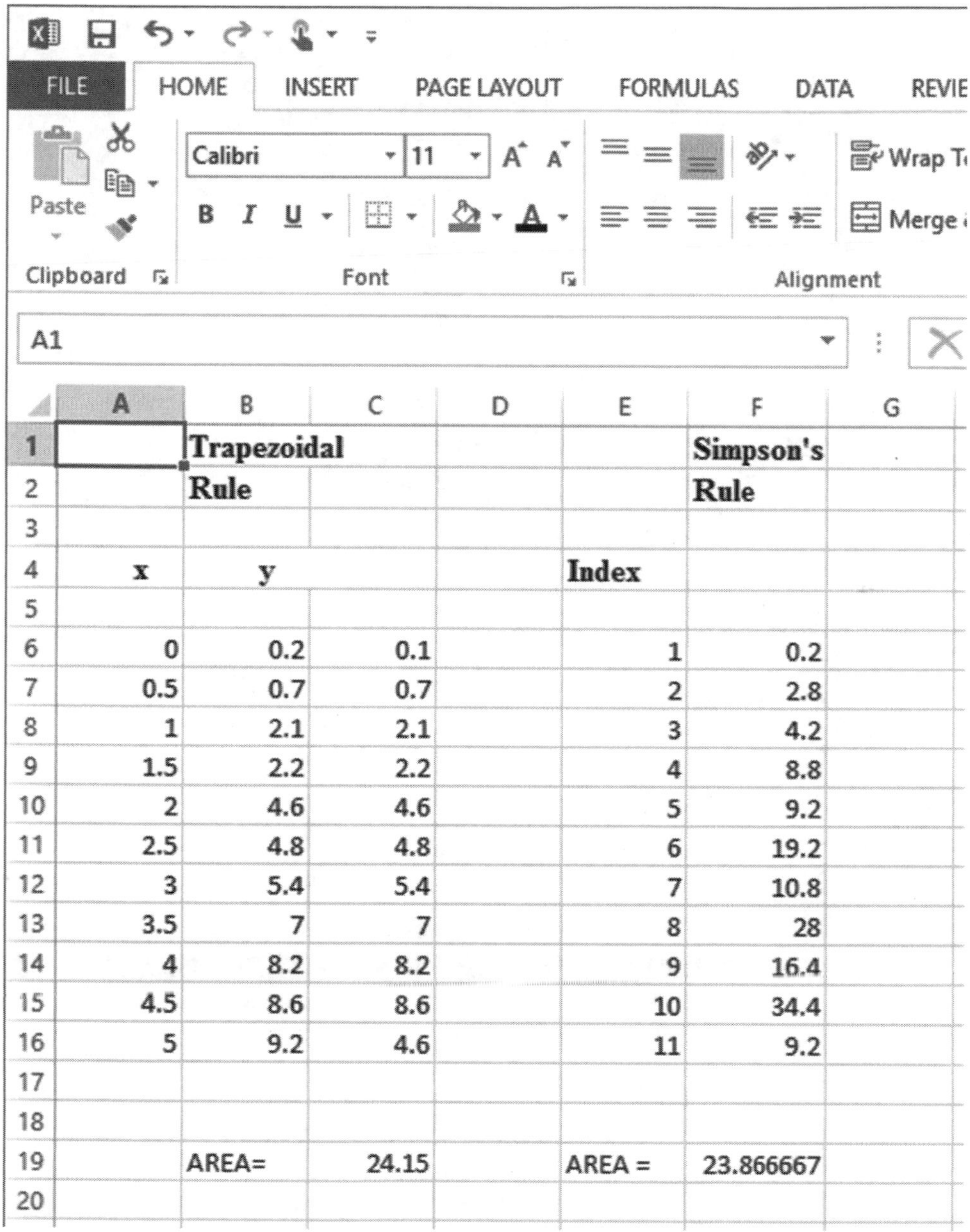

	A	B	C	D	E	F	G
1		Trapezoidal				Simpson's	
2		Rule				Rule	
3							
4	x	y			Index		
5							
6	0	0.2	0.1		1	0.2	
7	0.5	0.7	0.7		2	2.8	
8	1	2.1	2.1		3	4.2	
9	1.5	2.2	2.2		4	8.8	
10	2	4.6	4.6		5	9.2	
11	2.5	4.8	4.8		6	19.2	
12	3	5.4	5.4		7	10.8	
13	3.5	7	7		8	28	
14	4	8.2	8.2		9	16.4	
15	4.5	8.6	8.6		10	34.4	
16	5	9.2	4.6		11	9.2	
17							
18							
19		AREA=	24.15		AREA =	23.866667	
20							

FIGURE 6.6

format in the block at the bottom of the sheet, along with restrictions on their range of applicability in terms of the Re_L parameter. The equations are written in Excel format in cells C14, C15, and C16. The value of the Re parameter (Reynolds number) that determines which equation is to be used is calculated in cell C11. A written description of the actions called by the nested IF statements in cell C17 is given as follows:

The first (outer) IF statement is

If the value in C11 is <5E5 (true test result), go to cell C14 for the calculation.
If the value in C11 is not <5E5 (false test result), go to the next (inner) IF statement.

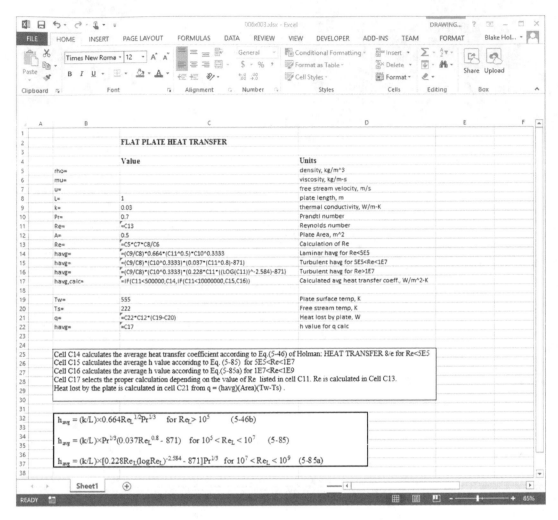

FIGURE 6.7

The second (inner) IF statement is

> If the value of C11 is <1E7 (true test result) [and also >5E5 from the first IF statement], go to cell C15 for the calculation.
> If the value in C11 is not <1E7 (false test result), go to cell C16 for the calculation.

Example 6.4: Use of IF Statement to Return Comments

The logical IF statement may be used to return a written comment instead of a calculation action as described in Example 6.3. In Figure 6.8, the numerical value in cell A3 is tested. If the value is >3, the comment YES is displayed in cell A4. If the value in cell A3 is <3 (i.e., the logical test is false), the comment NO is displayed in cell A4. The IF statement formula is shown along with two specific cases—for A3=2 (<3, false result), NO is displayed in A4; and for A3=4 (>3, true result), YES is displayed in A4. If long comments are to be displayed, it is probably better to have a short reference comment, such as "See text box number 1," etc., as the output of the IF statement.

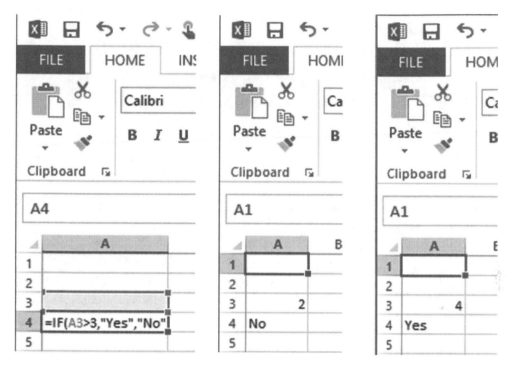

FIGURE 6.8

6.4 Histograms and Cumulative Frequency Distributions

Histograms are, very simply, a way of compartmentalizing various sets of data. Thus, we might construct a histogram that listed new cars in *bins*, or compartments according to their base Manufacturer's Suggested Retail Price (MSRP) prices, or their weight, or their engine displacement. A histogram for a digital camera provides a visual display that shows the distribution of light across the visual spectrum. Excel provides a means for arranging data in such histograms.

Example 6.5: Student Test Scores

The use of Excel for producing histograms and cumulative frequency distributions for a set of data will be demonstrated by analyzing a hypothetical set of student scores. These scores are shown in column B of Figure 6.9, along with a student-identifying number in column A. We wish to display the distribution of grades in 10-point ranges (or bins) of 90–100, 80–90, 70–80, 60–70, 50–60, and below 50. We specify the upper bound of each range (bin) as shown in cells D3:D7. The selection of 79.9, 89.9, etc. forces grades such as 80 or 90 to fall in the next higher bin.

In Excel 2016, creating a histogram is as easy as creating any other chart or graph. Using the data in column B of Figure 6.9, select the scores and insert a histogram chart in the worksheet. Excel will automatically create bins based on the data. The result of this default behavior is displayed in Figure 6.10.

To edit the bins displayed on the Histogram, right click on the horizontal axis and select the Format Axis menu option. Adjust the bin parameters in the formatting window displayed on the right side of the screen. Figure 6.11 displays the bins adjusted for the data in Figure 6.9 along with an updated Histogram.

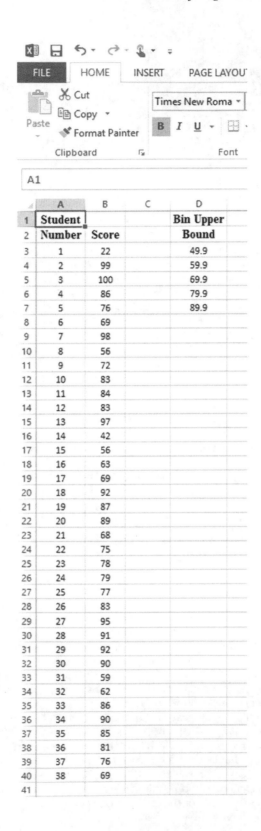

FIGURE 6.9

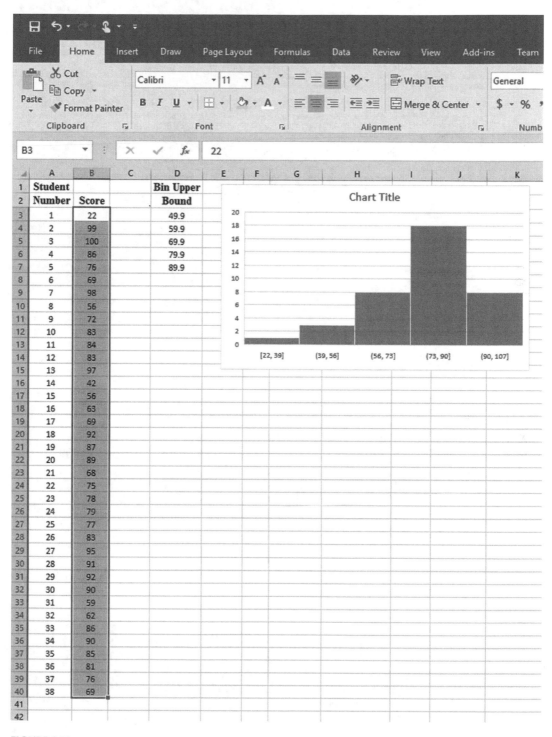

FIGURE 6.10

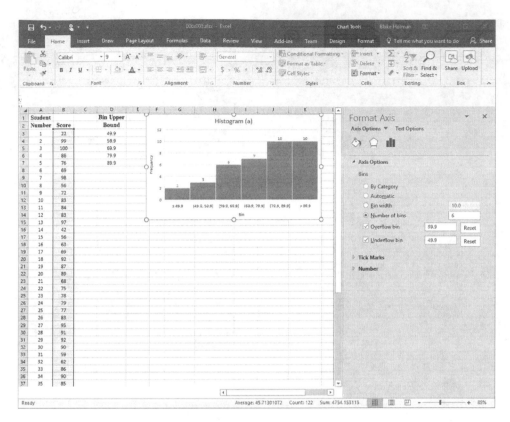

FIGURE 6.11

Cumulative frequency information is useful in this example, as are other statistical calculations. Cumulative frequency information is presented in cells F3:H9, and other statistical calculations are presented in cells J4:K9. Standard Excel functions were employed for most of these statistical calculations as shown in the display of the formulas in Figure 6.12. STDEV calculates the sample standard deviation, whereas STDEVP calculates the population standard deviation. They are nearly equal for 38 data points. An alternative way to calculate the arithmetic mean is to use the Excel AVERAGE function. Cell K4 would then become =AVERAGE(B3:B40).

In this example, the scores are weighted toward the higher grades. It must be a good class!

J	K
Other	**Parameters**
(calculated)	
Mean=	=(SUM(B3:B40))/38
STDEV=	=STDEV(B3:B40)
STDEVP=	=STDEVP(B3:B40)
Median=	=MEDIAN(B3:B40)
Max=	=MAX(B3:B40)
Min=	=MIN(B3:B40)

FIGURE 6.12

6.5 Normal Error Distributions

The Gaussian or normal error distribution is defined by

$$\partial(x, x_m, \sigma) = (1/2\pi\sigma^2)^{1/2} \times \mathrm{EXP}\left[-(x-x_m)^2/2\sigma^2\right] \tag{6.5}$$

where
 x=value for evaluation of function
 x_m=arithmetic mean of the distribution=SUM$(x_i)/n$ or calculated with the function
       AVERAGE$(x_1, x_2,\ldots, x_n)$
 n=number of x-values
 σ=standard deviation for the set of x-values

and ∂ is called the *probability density function.* The function is normalized so that

$$\int_{-\infty}^{+\infty} \partial(x)\,dx = 1.0 \quad \text{for } -\infty < x < +\infty \tag{6.6}$$

The *cumulative frequency* is defined by

$$\int_{-\infty}^{x} \partial(x)\,dx \text{ for limits from } -\infty \text{ to } x \tag{6.7}$$

The normal distribution function is called with

NORMDIST (x, mean, standard dev, cumulative)

where "mean" is the arithmetic mean of the set of data, and "standard dev" is the standard deviation for the data. For "cumulative," the word "true" is entered for a calculation of the cumulative frequency according to Equation 6.7, whereas the word "false" is entered to return a computation of the probability density according to Equation 6.5. For the special case of x_m=0 and standard dev=1.0, the function returns the value of the standard normal distribution. That same distribution may be called with the function

NORMSDIST (x, cumulative)

The inverse function

NORMINV (probability, mean, standard dev)

returns the value of x corresponding to the arguments in the function syntax.

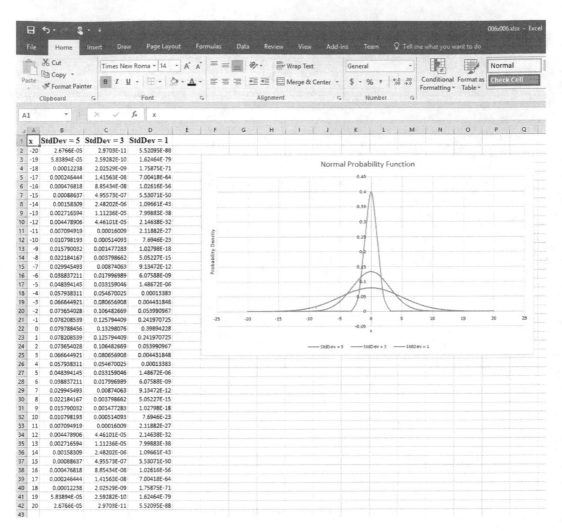

FIGURE 6.13

Example 6.6: Application of Normal Distribution and Inverse Functions

For practice purposes, some quick examples of applying the normal distribution and inverse functions are given as

$$\text{NORMDIST}(5, 0, 3, \text{FALSE}) = 0.0807$$

$$\text{NORMDIST}(5, 0, 3, \text{TRUE}) = 0.9522$$

$$\text{NORMDIST}(2, 0, 1, \text{TRUE}) = 0.9772 = \text{NORMDIST}(2)$$

$$\text{NORMINV}(0.9772, 0, 1) = 2$$

The calculations and graphs in Figures 6.13 and 6.14 show the density function and cumulative frequency for an arithmetic mean $x_m = 0$ and standard deviations of 1, 3, and 5. The density functions exhibit the familiar bell curve distribution that results

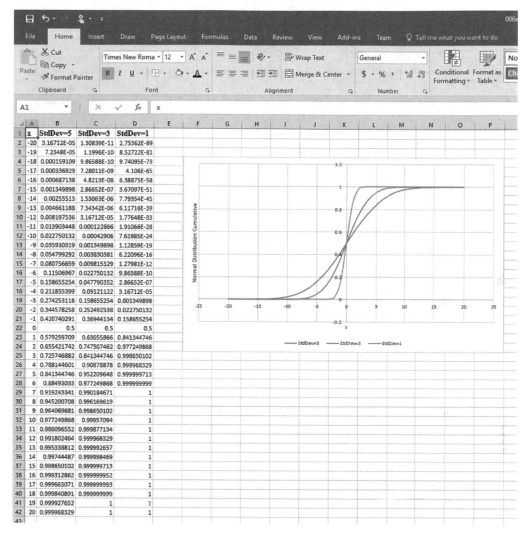

FIGURE 6.14

when a large number of samples are taken. The tabulation and plots are made for the range $-20 < x < +20$. The density function reaches a near zero value by $x = \pm 15$.

The chart in Figure 6.14 showing the cumulative frequency plot indicates that, for all standard deviation values, half the values lie above and half lie below the arithmetic mean. For a sufficiently large value of x, all cumulative frequency distribution curves approach a value of 1.0, i.e., all of the values of x are included.

The next example compares the cumulative frequency of the student scores examined in Example 6.5 with that of a normal distribution having the same mean and standard deviation values.

Example 6.7: Comparison of Actual and Normal Cumulative Frequencies

We have seen how a cumulative frequency plot for a set of data may be viewed in different perspectives by changing the boundaries of the bins on a histogram. The resulting histograms may or may not resemble a normal distribution. Comparison with a normal distribution may be made by taking uniform increments in x and a computation of the normal distribution having the same mean and standard deviation values as those of the data. For this comparison, we will use the set of student scores discussed previously.

The worksheet shown in Figure 6.13 lists the scale for the student scores shown in Example 6.5 in column A using two-point increments. The number of student scores up to each point is listed in column D. The cumulative frequency for the student scores is calculated in column B with =D3/38 because the total number of scores is 38. These entries stop when the score reaches 100. The cumulative frequency for a normal distribution with the same mean and standard deviation values is computed in column C using mean=77.868 and standard deviation=16.409. The function is then

$$= NORMDIST(x, mean, standard\ dev, cumulative)$$

$$= NORMDIST(x, 72.868, 16.409)$$

where x is the student score in column A. The functions are dragged-copied for the number of rows needed. For the student scores, the calculation is stopped when a score of 100 is reached, whereas the calculation for the normal distribution is carried past a score of 100 to approach a maximum frequency of 1.0. A type 5 scatter chart is employed for the graphical presentation shown in Figure 6.15. The comparison is not bad.

6.6 Calculation of Uncertainty Propagation in Experimental Results

A common calculation in the analysis of experimental data involves the propagation of uncertainties from the basic measured parameters into a result calculated from these variables. The experimentalist must set the values of the uncertainties in the primary measurements based on calibration information, manufacturer's specifications for the instruments, and the overall laboratory experience. An experimental variable will then be expressed as

$$x_1 = x_1 \pm w_{x1}$$

where w_{x1} is the uncertainty in x_1.

The calculated result y is expressed in terms of measured parameters through a known function ∂. (It must be known to calculate a result!) Thus,

$$y = \partial(x_1, x_2, \ldots, x_n) \tag{6.8}$$

If the uncertainties in the experimental x-values are all expressed with the same probabilities, the uncertainty in y may be calculated with

$$w_y = [(\partial f/\partial x_1)^2 w_{x1}^2 + (\partial f/\partial x_2)^2 w_{x2}^2 + \cdots]1/2 \tag{6.9}$$

An alternative to performing the partial differentiations is a calculation of the derivatives using finite-difference approximations. Thus, the partial derivatives are approximated by

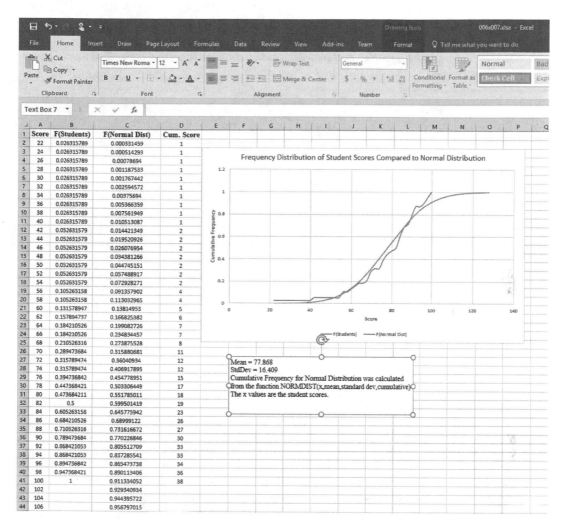

FIGURE 6.15

$$(\partial f / \partial x_1) \approx [f(x_1 + \Delta x_1, x_2, \ldots) - f(x_1, x_2, \ldots)] / \Delta x_1 \qquad (6.10)$$

where $\Delta x_1 \approx 0.01 x_1$ or some small increment in x_1.

Example 6.8: Uncertainty Calculation for Three Variables

The Excel procedure for making this calculation will be demonstrated by application to a rather simple ∂ function. The procedure is easily extended to more complicated functions.

We assume that y is a function of three experimental variables through the relation

$$y = x_1^2 x_2 x_3 \qquad (6.11)$$

The worksheet is set up as shown in Figure 6.16, with the numerical values of x to be entered at B1:B3 and their uncertainties at D1:D3. If the problem involves more

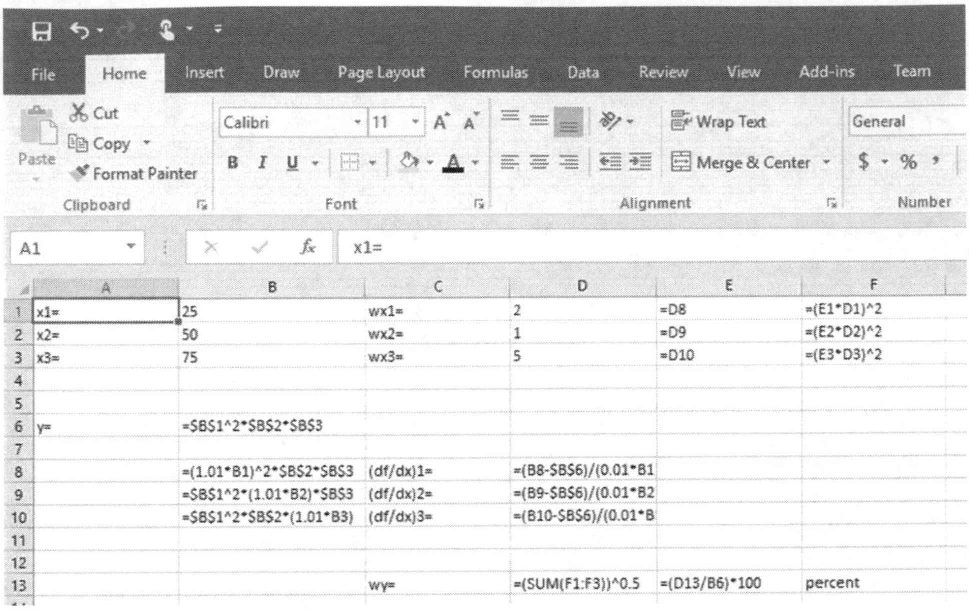

	A	B	C	D	E	F
1	x1=	25	wx1=	2	=D8	=(E1*D1)^2
2	x2=	50	wx2=	1	=D9	=(E2*D2)^2
3	x3=	75	wx3=	5	=D10	=(E3*D3)^2
4						
5						
6	y=	=$B$1^2*$B$2*$B$3				
7						
8		=(1.01*B1)^2*$B$2*$B$3	(df/dx)1=	=(B8-$B$6)/(0.01*B1		
9		=$B$1^2*(1.01*B2)*$B$3	(df/dx)2=	=(B9-$B$6)/(0.01*B2		
10		=$B$1^2*$B$2*(1.01*B3)	(df/dx)3=	=(B10-$B$6)/(0.01*B		
11						
12						
13			wy=	=(SUM(F1:F3))^0.5	=(D13/B6)*100	percent

FIGURE 6.16

parameters, there would be correspondingly more rows to list the experimental values and associated uncertainties.

The formula for the y function is written at B6 according to Equation 6.11, using absolute cell references. The formula is then copied to B8:B10, and $B$1 is replaced by 1.01*B1 in the first formula, $B$2 is replaced by 1.01*B2 in the second formula, and $B$3 is replaced by 1.01*B3 in the third formula. Note that Δx is taken as 0.01x in all cases. This increment should be small enough to give satisfactory results.

The values of the $\partial f/\partial x$ are calculated at D8:D10 using Equation 6.10. Again, the equations may be dragged-copied for complicated functions with many variables.

The values of the $\partial f/\partial x$ are copied to E1:E3 for display and convenience purposes, and the values of the $(\partial f/\partial x)^2 w_x^2$ are computed at F1:F3. Again, dragging-copying of the formulas may be used when many variables are involved.

Finally, w_y is computed with Equation 6.9 at D13, and the percentage uncertainty is calculated at cell E13.

The results for numerical values of

$$x_1 = 25;\ x_2 = 50;\ x_3 = 75$$

and

$$w_{x1} = 2(8\%);\ w_{x2} = 1(2\%);\ w_{x3} = 5(6.7\%)$$

are shown in Figure 6.17.

Comment

By observing the relative values of the entries at F1:F3, the experimentalist can usually obtain an idea of the variables that cause the greatest uncertainty in the calculated result. In this example,

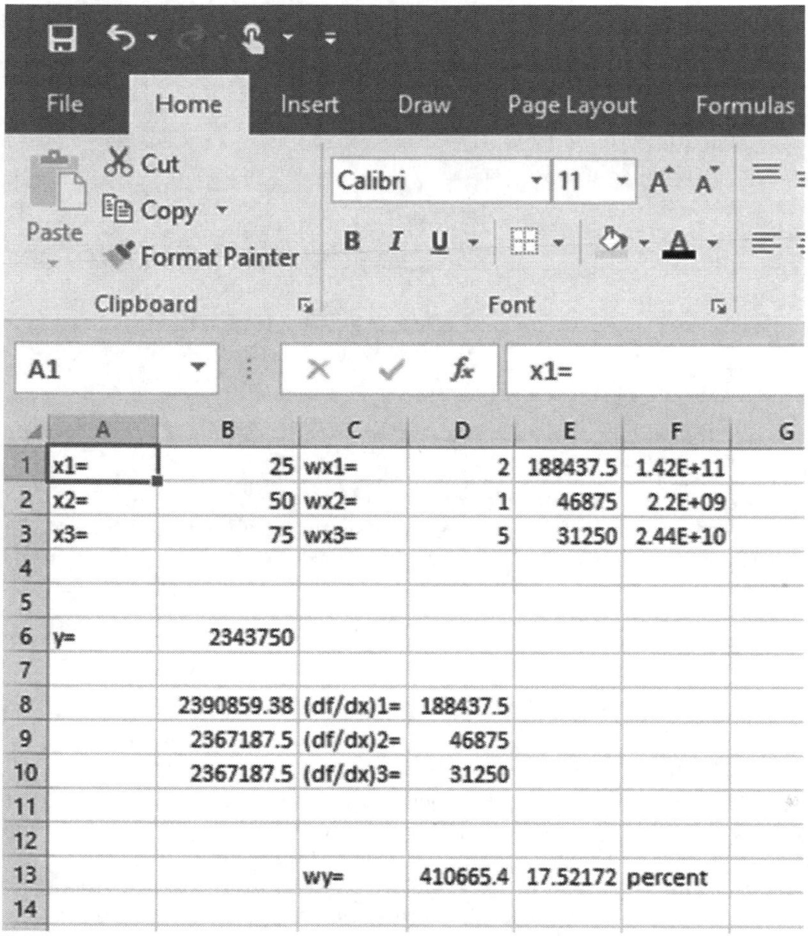

	A	B	C	D	E	F	G
1	x1=	25	wx1=	2	188437.5	1.42E+11	
2	x2=	50	wx2=	1	46875	2.2E+09	
3	x3=	75	wx3=	5	31250	2.44E+10	
4							
5							
6	y=	2343750					
7							
8		2390859.38	(df/dx)1=	188437.5			
9		2367187.5	(df/dx)2=	46875			
10		2367187.5	(df/dx)3=	31250			
11							
12							
13			wy=	410665.4	17.52172	percent	
14							

FIGURE 6.17

$$F1 = 1.4 \times 10^{11}$$

$$F2 = 2.2 \times 10^{9}$$

$$F3 = 2.4 \times 10^{10}$$

Clearly, variable x_1 causes the greatest contribution to the uncertainty of the result. Not only does it have the greatest percentage uncertainty to start with, but it also influences the calculation of the result through a square term. From the standpoint of experiment planning, the instrumentation or experimental technique associated with x_1 should be improved first to effect the greatest reduction in uncertainty of the results. In contrast, variable x_2 contributes little to the uncertainty of the result.

Example 6.9: Uncertainty in Measurement of Heat-Transfer Coefficient

Forced-convection heat-transfer coefficients may be measured by observing the temperature rise of a fluid as it passes through the convection channel. The heat transfer may be expressed by

$$q = mc(T_2 - T_1) \tag{6.12}$$

where
  q=heat-transfer rate, W
  m=mass-flow rate of fluid, kg/s
  c=specific heat of the fluid, J/kg °C
  T_1=entering fluid temperature
  T_2=exiting fluid temperature
This heat transfer may also be expressed in terms of the convection heat-transfer coefficient h with

$$q = hA(T_w - T_{avg\,fluid}) \tag{6.13}$$

where
  h=heat-transfer coefficient, W/m$^{2\,°C}$
  A=surface area, m^2
  T_w=wall or surface temperature in contact with the fluid
The average fluid temperature may be taken as the mean between the entering and exiting fluid temperatures. Setting the two expressions for heat transfer equal, h may be expressed as

$$h = mc(T_2 - T_1)/A\left[T_w - (T_2 + T_1)/2\right] \tag{6.14}$$

We now examine an experimental situation in which the uncertainties in the area, specific heat, and mass flow are assumed to be small in relation to the uncertainties in the temperature measurements. Accordingly, we will set their uncertainties to zero in the analysis of the overall uncertainty of the determination of h, the heat-transfer coefficient.

The worksheet is set up as shown in Figure 6.18, with A=1, m=1, c=4180 (the value for water), T_1=55°C, T_2=65°C, and T_w=85°C. The uncertainties in each of the temperature determinations are assumed to be ±1.0°C (<2%). In this example, we will see how the temperature levels greatly influence the uncertainty in h even when the uncertainties in the individual temperatures remain constant.

	A	B	C		D	E	F
1	T1=	55	wT1=		1	=D11	=(E1*D1)^2
2	T2=	65	wT2=		1	=D12	=(E2*D2)^2
3	Tw=	85	wTw=		1	=D13	=(E3*D3)^2
4	A=	1	wA=		0		
5	m=	1	wm=		0		
6	c=	4180	wc=		0		
7							
8							
9	h=	=$B$5*$B$6*($B$2-$B$1)/($B					
10							
11		=$B$5*$B$6*($B$2-1.01*B:	dh/dT1=		=(B11-B9)/(0.01*B1)		
12		=$B$5*$B$6*(1.01*$B$2-B:	dh/dT2=		=(B12-B9)/(0.01*B2)		
13		=$B$5*$B$6*($B$2-$B$1)/($B	dh/dTw=		=(B13-B9)/(0.01*B3)		
14							
15			wh=		=(SUM(F1:F3))^0.5	=(D15/B9)*100	percent
16							

FIGURE 6.18

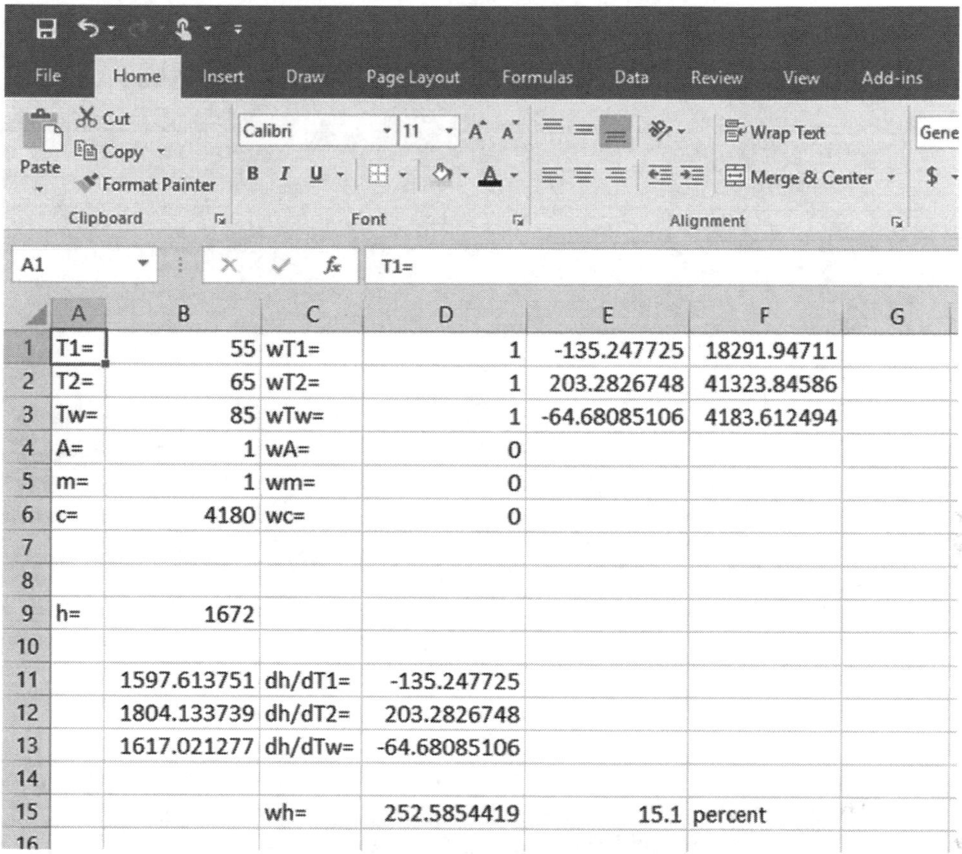

	A	B	C	D	E	F	G
1	T1=	55	wT1=	1	-135.247725	18291.94711	
2	T2=	65	wT2=	1	203.2826748	41323.84586	
3	Tw=	85	wTw=	1	-64.68085106	4183.612494	
4	A=	1	wA=	0			
5	m=	1	wm=	0			
6	c=	4180	wc=	0			
7							
8							
9	h=	1672					
10							
11		1597.613751	dh/dT1=	-135.247725			
12		1804.133739	dh/dT2=	203.2826748			
13		1617.021277	dh/dTw=	-64.68085106			
14							
15			wh=	252.5854419	15.1	percent	
16							

FIGURE 6.19

As in Example 6.8, we take the finite-difference change in each variable as 0.01 of the variable for approximating the partial derivatives in cells B11:B13 and D11:D13. Four sets of results are shown as follows. For the given set of temperature data, the results are shown in Figure 6.19, having an uncertainty in h of 15.1% of the calculated value of 1672 W/m^2°C. In Figure 6.20 the value of T_1 is reduced to 40°C, while keeping the other two temperatures the same. The resulting uncertainty in h is now only 6.8%, or less than half the previous result. In Figure 6.21, the inlet and exit temperatures are returned to their original values, but the wall temperature is lowered to 70°C. The uncertainty in h increases to 18.8%. Finally, in Figure 6.22, the inlet temperature is increased to 60°C from 55°C, while T_2 and T_w remain at their original values. In this case, the uncertainty in h increases to the large value of 29.2%. Note that the uncertainty in each temperature has remained the same for all four calculations, i.e., ±1°C.

The reason there is so much variation in the four results is that the uncertainties in the individual temperatures produce larger or smaller uncertainties in the temperature difference calculations of (T_2-T_1) and $(T_w-T_{avg\ fluid})$ that appear in the formula for h, depending on the values of the temperatures themselves. If a fifth calculation were made with T_1 raised to 63°C while retaining $T_2=65$°C and $T_w=85$°C, the resulting uncertainty in h would be a whopping 72%.

Uncertainties in A, c, and m could also be included in the calculation, but it should be obvious from the aforementioned results that they would have relatively less influence on the final uncertainty in h.

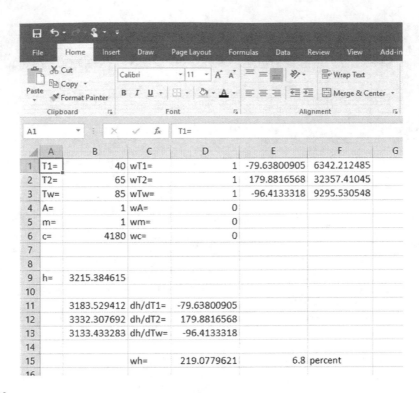

FIGURE 6.20

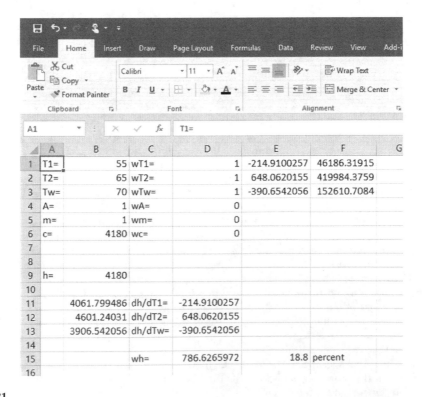

FIGURE 6.21

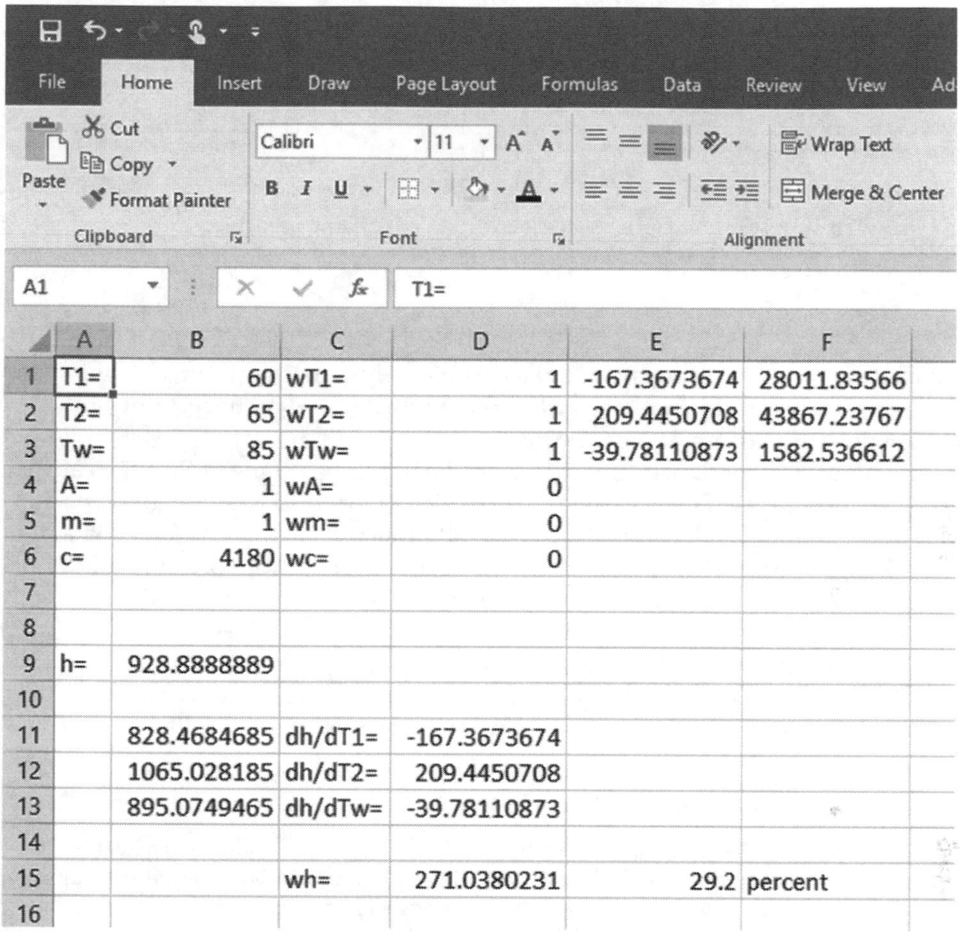

	A	B	C	D	E	F
1	T1=	60	wT1=	1	-167.3673674	28011.83566
2	T2=	65	wT2=	1	209.4450708	43867.23767
3	Tw=	85	wTw=	1	-39.78110873	1582.536612
4	A=	1	wA=	0		
5	m=	1	wm=	0		
6	c=	4180	wc=	0		
7						
8						
9	h=	928.8888889				
10						
11		828.4684685	dh/dT1=	-167.3673674		
12		1065.028185	dh/dT2=	209.4450708		
13		895.0749465	dh/dTw=	-39.78110873		
14						
15			wh=	271.0380231	29.2	percent
16						

FIGURE 6.22

6.7 Fractional Uncertainties for Product Functions of Primary Variables

Consider the case in which the result y is expressed as a product function of the primary measurement variables x_i, each raised to a power a_i. Thus,

$$y = x_1^{a_1} x_2^{a_2},\ldots,x_n^{a_n} \tag{6.15}$$

Evaluating the partial derivatives in Equation 6.9 and dividing by y from Equation 6.15 gives

$$w_y/y = \left[\Sigma(a_i w_{xi}/x_i)^2\right]^{1/2} \tag{6.16}$$

When sums or differences of variables are involved in the result along with power variables, some variable substitutions may be made to fit the result function into the form of Equation 6.15. For example, the relation

$$h = q/A(T_1 - T_2) \tag{6.17}$$

could be rewritten with a variable substitution $\Delta T = T_1 - T_2$ to read

$$h = q/A\Delta T \tag{6.18}$$

The uncertainty in ΔT could be evaluated separately and then Equation 6.18 could be treated as a power relation with $a_q = 1$, $a_A = -1$, and $a_{\Delta T} = -1$. The uncertainty in the result could then be evaluated using Equation 6.16.

For the product function cases, an Excel worksheet may be arranged to either calculate the fractional uncertainty in the result directly from Equation 6.16 or compute the influence coefficients $(a_i w_{xi}/x_i)^2$ separately before the individual influence coefficients are used in the final result. Although the latter calculation is a bit longer, it offers the opportunity to examine the relative magnitudes of the influence coefficients and, thus, to obtain additional insight into the behavior of the experimental results. Note that a calculation using Equation 6.16 does not involve any finite-difference approximations for the partial derivatives. Depending on the complexity of the experiment and the form of the result function, some combination of methods may be chosen for the calculation, as described in Example 6.10.

Example 6.10: Modification of Example 6.9 for Uncertainties in m and c

Example 6.9 assumed exact determination of the mass flow m and specific heat c. Note that both variables appear as a product function in Equation 6.14. Now, let us examine the case in which m and c are not exact but have associated uncertainties of wm and wc. In addition, we assume that the mass-flow rate is determined from a flow measurement in which

$$m = C_D A_f \left(p\Delta p\right)^{1/2} \tag{6.19}$$

where C_D is a flow coefficient with a known uncertainty w_{Cd}, A_f is an area known almost exactly, p is a pressure measurement, and Δp is a differential pressure measurement. Equation 6.19 is a product function, and w_m/m can be calculated using Equation 6.16, along with a determination of w_c/c. A variable f(T) may be defined by

$$f(T) = (T_2 - T_1)/[T_w - (T_2 + T_1)/2] \tag{6.20}$$

and a separate determination of $w_{f(T)}/f(T)$ made using the method of Example 6.9. Then, Equation 6.14 is expressed as a product result function

$$h = mc[f(T)]$$

with

$$am = 1, a_c = -1, \text{ and } a_{f(T)} = 1$$

A direct application of Equation 6.16 can be made to calculate the fractional uncertainty w_h/h. Extension of this technique to more complex result functions is straightforward but may be tedious in execution. A discussion of the importance of uncertainty determinations in experiment design and various types of data analyses is given in Reference 2.

Example 6.11: Application of Product Method to Example 6.8

The y function given in Equation 6.11 is a product function with $a_{x1}=2$, $a_{x2}=1$, and $a_{x3}=1$. From Example 6.8, we have

$$x_1 = 25; x_2 = 50; x_3 = 75; w_{x1} = 2; w_{x2} = 1; w_{x3} = 5$$

The result for y is

$$y = (25)^2(50)(75) = 2.344 \times 10^6$$

Application of Equation 6.16 gives the fractional uncertainty for y as

$$w_y/y = \left\{ [(2)(2)/25]^2 + [(1)(1)/50]^2 + [(1)(5)/75]^2 \right\}^{1/2} = 0.17448$$

This value may be compared with the result given in Example 6.8 as 0.17522. The 0.17448 value is the correct one because it does not rely on a finite-difference approximation to the partial derivatives. Even so, the two results for fractional uncertainty agree within 0.5%. Clearly, a worksheet could be constructed to examine the relative effects of the influence coefficients on the result, similar to that in Example 6.8.

6.8 Multivariable Linear Regression

Linear regression analysis with a single independent variable may be performed using a graphical display and the trendline and R^2 features discussed in Sections 3.7 and 3.8.

Excel may also be used to perform a least-squares regression calculation for a linear function with multiple independent variables.

Consider the variable y expressed as the linear function:

$$y = m_1 x_1 + m_2 x_2 + \cdots + m_n x_n + b \tag{6.21}$$

which becomes the familiar linear equation for a straight line,

$$y = mx + b \tag{6.22}$$

when only a single independent variable is involved. Excel executes the least-squares linear regression calculation using the worksheet function LINEST, which requires the following syntax:

$$\text{LINEST}(\text{y-values}, x_n\text{-values}, \text{constant}(\text{true/false}), \text{statistics}(\text{true/false})) \qquad (6.23)$$

where
   n is the number of independent variables in Equation 6.21
   y-values is a known array of numerical y-values that may equal or exceed $n+1$
   x_n-values is a known array of numerical values for the independent variables that may equal or exceed $n+1$ for each variable
   constant(true/false) is a logical statement; TRUE is entered if the regression analysis is to include a determination of the constant b in Equation 6.21; FALSE is entered if the value of b is to be taken as zero
   statistics(true/false) is a logical statement; TRUE returns a calculation of certain statistical functions as described in the following text; entry of FALSE returns no statistical calculations
   The LINEST function must be entered as an array. A block of cells is first selected (activated), the function is typed in the upper-left corner of the block, and the key combination Ctrl+Shift+Enter is pressed. The format of the results array is as follows:

Row					
1	m_n	m_{n-1}	...	m_1	b
2	se_n	se_{n-1}	...	se_1	se_b
3	R^2	se_y			
4	F	df			
5	ss_{reg}	$ss_{residual}$			

Thus, an array of 5 rows and $n+1$ columns should be selected for the presentation of the results. No other information may appear in this block of cells.
   The m and b values displayed are used in Equation 6.21, R^2 is the coefficient of determination described in Section 3.8, these terms are the standard errors for the various quantities, ss_{reg} is the regression sum of squares, $ss_{residual}$ is the residual sum of squares, df is the number of degrees of freedom, and F is a statistical function. The reader should consult Excel Help under the LINEST function and "standard errors" for a description of the standard error and statistical functions. Our primary concern here is to determine the m_n and b constants for the procedure.

Example 6.12: Linear Regression for Three Variables

Numerical values for y as a function of the three variables x1, x2, and x3 are shown in the worksheet of Figure 6.23, with eight numerical values for each variable. A block of cells for the output results of the LINEST function is selected as A12:D16. The LINEST function is then typed in cell A12 as shown in Figure 6.24, in accordance with the syntax of Equation 6.23. The array containing the numerical y-values is D2:D9, and the array containing all of the numerical x_n values is A2:C9. Entry of

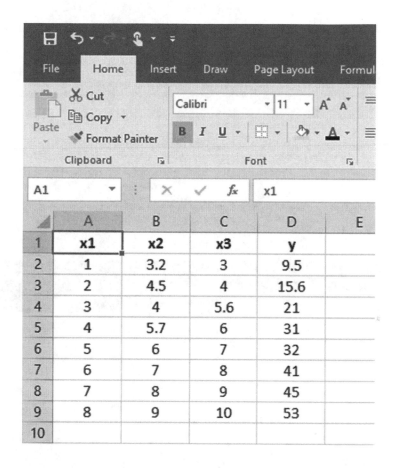

FIGURE 6.23

FIGURE 6.24

the first TRUE calls for a determination of the constant b, whereas entry of the second TRUE calls for display of the statistical results. Ctrl+Shift+Enter is then pressed to execute the function.

The results are shown in Figure 6.25 along with the original numerical data. In accordance with the output format described in Section 6.8, we have

$$m_1 = 3.247873$$

$$m_2 = 2.041483$$

$$m_3 = 1.241881$$

$$b = -3.86409$$

$$R_2 = 0.991044$$

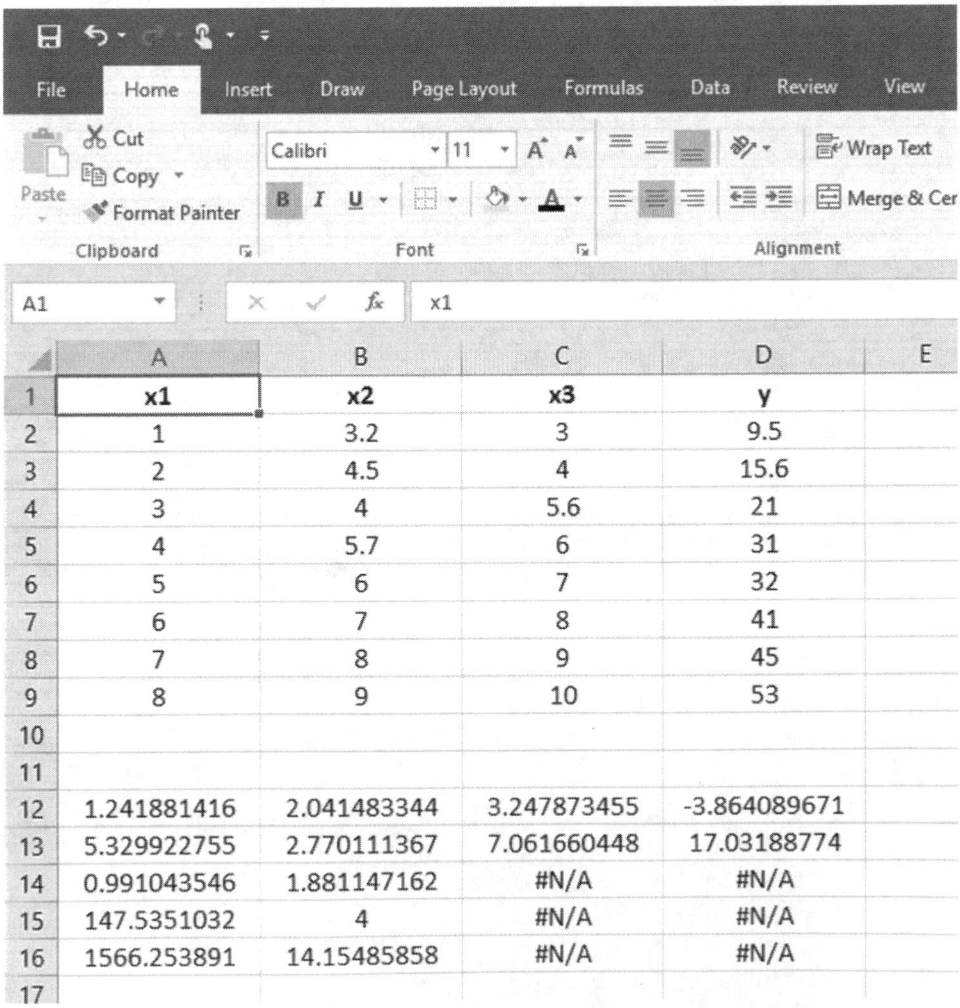

	A	B	C	D	E
1	x1	x2	x3	y	
2	1	3.2	3	9.5	
3	2	4.5	4	15.6	
4	3	4	5.6	21	
5	4	5.7	6	31	
6	5	6	7	32	
7	6	7	8	41	
8	7	8	9	45	
9	8	9	10	53	
10					
11					
12	1.241881416	2.041483344	3.247873455	-3.864089671	
13	5.329922755	2.770111367	7.061660448	17.03188774	
14	0.991043546	1.881147162	#N/A	#N/A	
15	147.5351032	4	#N/A	#N/A	
16	1566.253891	14.15485858	#N/A	#N/A	
17					

FIGURE 6.25

and the final regression formula for y is

$$y = 3.247873x_1 + 2.041483x_2 + 1.241881x_3 - 3.86409 \tag{6.24}$$

Note that the values of the m_n appear in reverse order to the columns for the numerical values of the x_n.

It is interesting to compare Equation 6.24 with the original numerical data. For the x-values in row 2, we may calculate

$$y = (3.247873)(1) + (2.041483)(3.2) + (1.241881)(3) - 3.86409$$
$$= 9.642172 \text{ as compared to a data value of 9.5 in cell D2} \tag{6.25}$$

For the x-values in row 9, we obtain

$$y = (3.247873)(8) + (2.041483)(9) + (1.241881)(10) - 3.86409$$
$$= 52.911 \text{ as compared to a data value of 53 in cell D9} \tag{6.26}$$

The value of $R^2 = 0.991$ indicates a fairly good regression result.

Example 6.13: Modeling Performance of an Air-Conditioning Unit

The cooling performance of a commercial air-conditioning unit depends on the indoor and outdoor temperatures. The appropriate indoor temperature that governs the cooling process is the wet bulb temperature, which we designate as T_{ew} (°F). The outdoor temperature that governs the cooling process is the dry bulb temperature that enters the conditioning unit, which we designate as T_c (°F). The cooling performance of the unit is stated in units of thousands of Btu/h and is designated as Q_{ew}.

The upper portion of Figure 6.26 gives the performance data for an actual air-conditioning unit as specified by the manufacturer. These data are based on the actual measurements of the unit's cooling capacity and are fitted to standard temperature ranges agreed to by the manufacturer. Although these data are useful as they stand for unit-sizing activities, an analytical expression of the data may be desired for modeling studies in conjunction with performance of other devices associated with a particular application, such as pumps and fans. The LINEST worksheet function is a suitable way to obtain such a relation.

The dependent variable or "y-function" is Q_{ew}, whereas the independent variables are T_{ew} and T_c. A group of cells (A21:C25) is activated as shown in Figure 6.26, and the LINEST function is entered in cell A21. The array for the y-variable is C4:C19, whereas the array containing the x-variables is A4:B19. TRUE is entered to obtain a fit with a constant value b, with TRUE also entered to obtain values of the statistical parameters. Ctrl+Shift+Enter is pressed, and the regression parameters are listed in cells A21:C21 as shown in Figure 6.27 with the values

$$m_{Tew} = 0.526$$

$$m_{Tc} = -0.228$$

$$b = 34.723$$

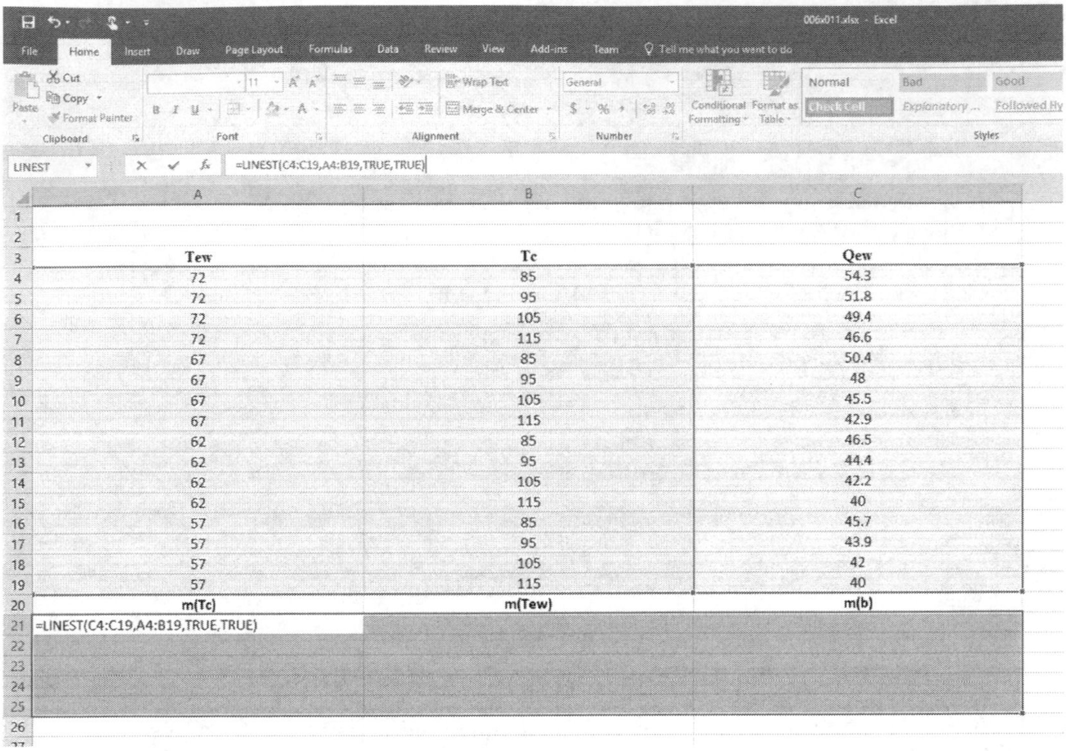

FIGURE 6.26

The resulting linear regression formula to be used for calculating Q_{ew} is, therefore,

$$Q_{ew} = 0.526 T_{ew} - 0.228 T_c + 34.723 \tag{6.27}$$

with $R^2 = 0.943$

Values calculated from this formula may be compared with the original data by calculating a two-variable table using the DATA/FORECAST/What-If-Analysis/Data Table command described in Section 2.15.

6.9 Multivariable Exponential Regression

In addition to the linear regression analysis with the LINEST worksheet function described in Section 6.8, Excel also provides an exponential regression operation, which employs the LOGEST worksheet function. This function may be used to obtain a least-squares regression to fit exponential functions of single or multiple independent variables, although single-variable regression is more easily accomplished as described in Sections 3.10 and 3.11. The general form of equation used for the fit is

$$y = b \times m_1^{x_1} \times m_2^{x_2} \times \cdots \times m_n^{x_n} \tag{6.28}$$

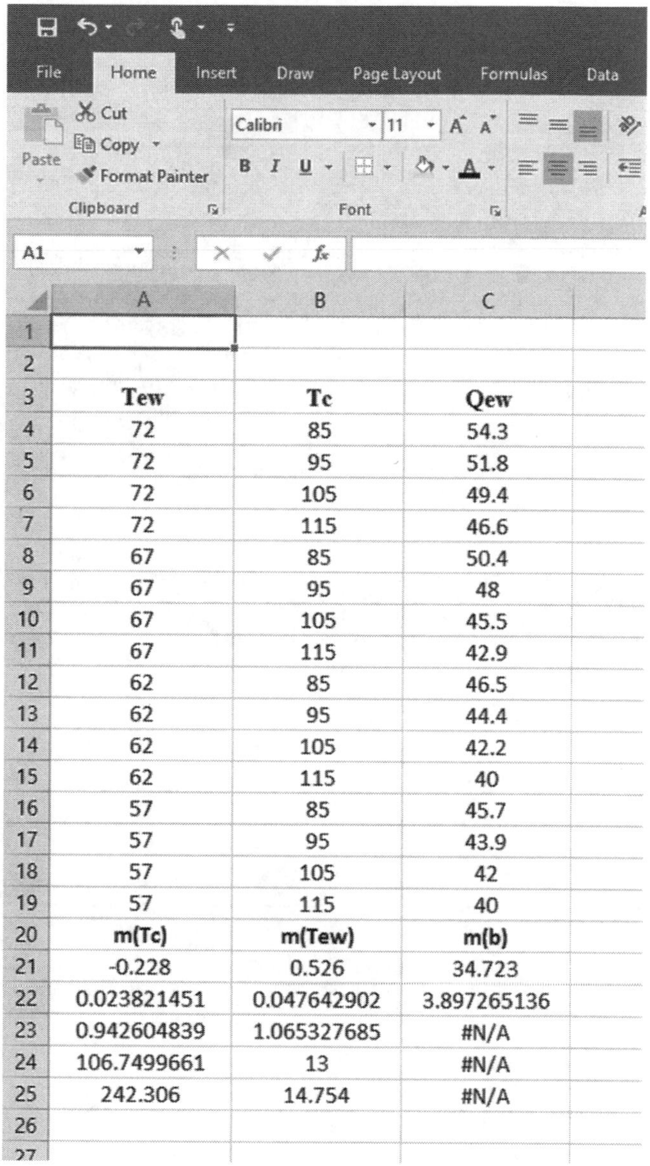

FIGURE 6.27

For a single variable, the equation becomes

$$y = b \times m^x \tag{6.29}$$

Execution of the least-squares regression is accomplished in the same manner as that for the linear analysis described in Section 6.8. Now, the LOGEST worksheet function is executed with the following syntax:

$$\text{LOGEST}(y\text{-values}, x_n\text{-values}, \text{constant(true/false)}, \text{statistics(true/false)}) \tag{6.30}$$

where the nomenclature is the same as that for Equation 6.23. As with LINEST, the LOGEST function must be entered as an array by (1) selecting cells for the output array (same format as in Section 6.8), (2) entering the equation for the LOGEST function in the upper-left cell of the array selected, and (3) pressing Ctrl+Shift+Enter. Use of the method is described in the following example.

Example 6.14: Multivariable Exponential Regression

As a matter of interest, we will perform a multivariable exponential regression on the same data set used in Example 6.13 for a linear analysis. The data are displayed as before in Figure 6.28. A block of cells is selected for the output array as shown in the right-side insert of this figure, and the LOGEST function is entered in cell A21. Ctrl+Shift+Enter is pressed, producing the output shown in cells A21:C25 in Figure 6.29. We have the following results:

$$m_{Tc} = 0.995039$$

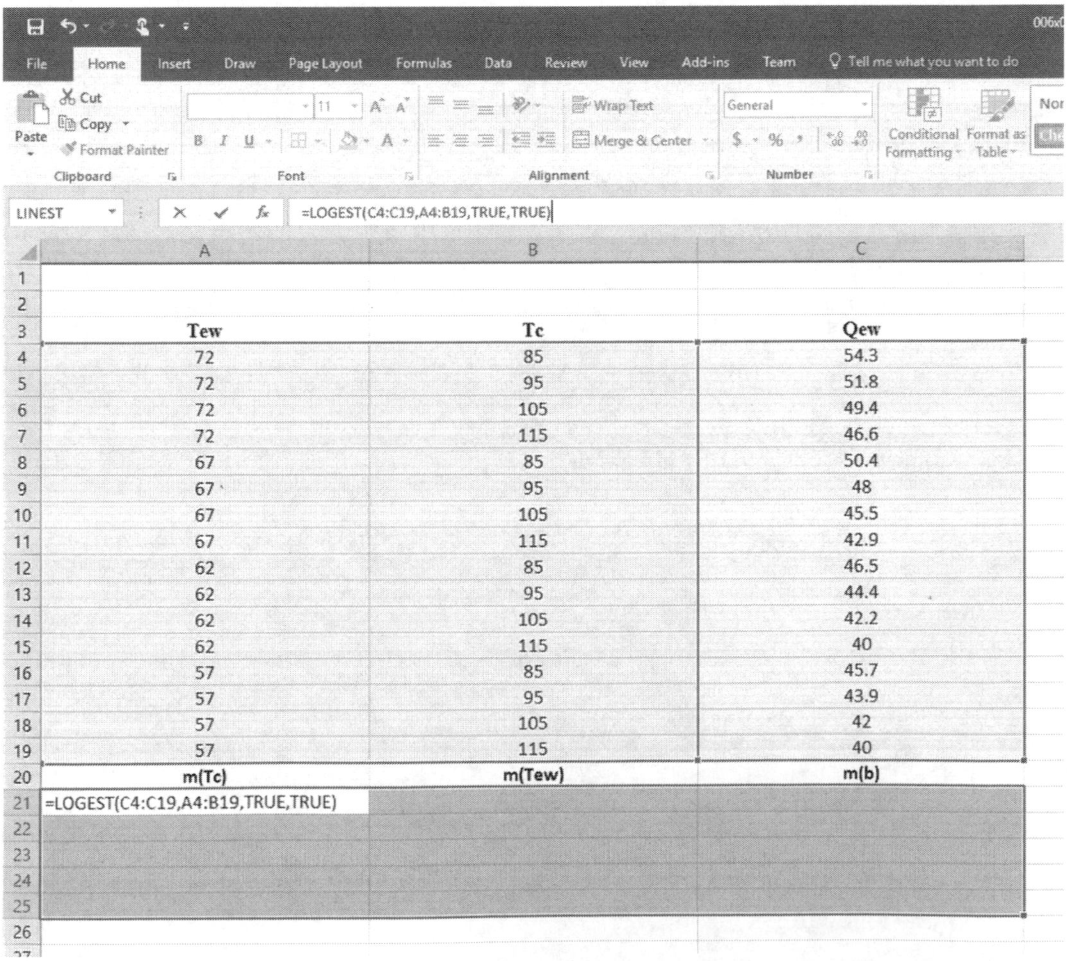

The formula bar reads: `=LOGEST(C4:C19,A4:B19,TRUE,TRUE)`

	A	B	C
1			
2			
3	**Tew**	**Tc**	**Qew**
4	72	85	54.3
5	72	95	51.8
6	72	105	49.4
7	72	115	46.6
8	67	85	50.4
9	67	95	48
10	67	105	45.5
11	67	115	42.9
12	62	85	46.5
13	62	95	44.4
14	62	105	42.2
15	62	115	40
16	57	85	45.7
17	57	95	43.9
18	57	105	42
19	57	115	40
20	**m(Tc)**	**m(Tew)**	**m(b)**
21	=LOGEST(C4:C19,A4:B19,TRUE,TRUE)		
22			
23			
24			
25			
26			

FIGURE 6.28

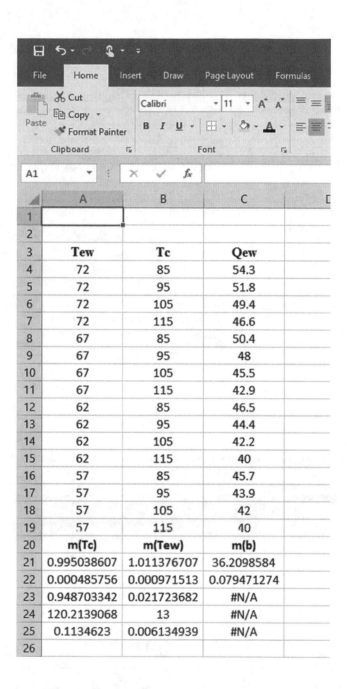

FIGURE 6.29

$$m_{Tew} = 1.011377$$

$$b = 36.20986$$

which produce the final equation for Q_{ew} as

$$Q_{ew} = 36.20986 \times 0.995039^{Tc} \times 1.011377^{Tew} \qquad (6.31)$$

Example 6.15: Combined Regression Analysis

An inspection of the data in Figure 6.26 suggests a linear variation of the cooling capacity Q_{ew} with the condenser temperature T_c. The variation with T_{ew} appears to be nonlinear. These facts suggest that a combination regression analysis using a nonlinear variation with T_{ew} may produce a better fit to the data points than the linear regression developed in Example 6.13.

First, we reconstruct the data table of Figure 6.26 in the format shown in Figure 6.30. Next, a chart is constructed for these data with T_{ew} as abscissa, and is displayed with the original data in Figure 6.31. The curves are obtained for each value of T_c. For each of these lines, a second-degree polynomial (quadratic) is fitted using the method described in Section 3.7. The resulting trend lines are plotted, and the trend line equations are displayed alongside the graph. The average value of the coefficient of x^2 (T_{ew}^2) is 0.034, and the average value of the coefficient of x (T_{ew}) is −3.925. The apparent linear variation of Q_{ew} with T_c suggests that a new variable $Q_{ew'}$ may be defined as

$$Q_{ew'} = Q_{ew} - 0.034T_{ew}^2 + 3.925T_{ew} = mT_c + b \qquad (6.32)$$

where m and b are constants to be determined using a linear regression analysis of $Q_{ew'}$ vs. T_c. The values of $Q_{ew'}$ are computed by using the DATA/FORECAST/What-If-Analysis/Data Table command described in Section 2.15, and the results are shown in Figure 6.32.

The resulting values of $Q_{ew'}$ are then plotted as a function of T_c as shown in Figure 6.33. A linear trendline is plotted through the data points, and the resulting trendline equation is displayed (with T_c substituted for x) as:

$$Q_{ew'} = -0.2302T_c + 179.51 \qquad (6.33)$$

Equations 6.32 and 6.33 are combined to give the final correlation for Q_{ew} as:

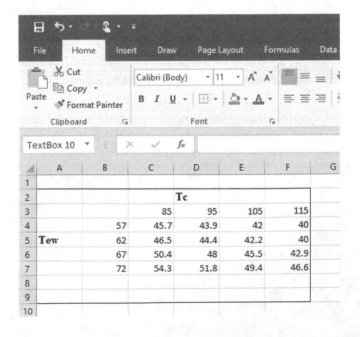

FIGURE 6.30

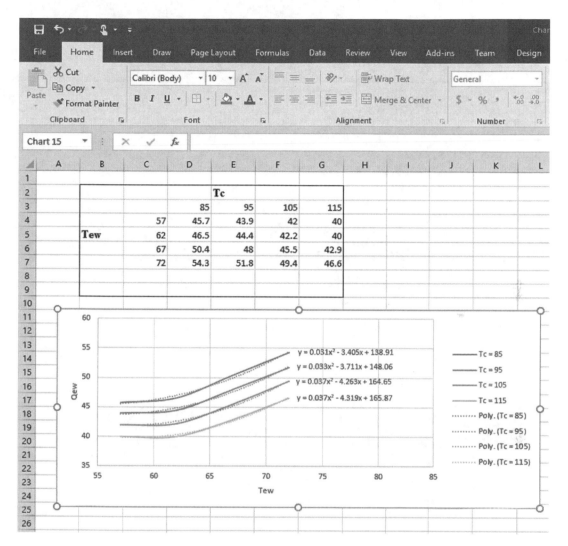

FIGURE 6.31

$$Q_{ew} = 0.034T_{ew}^{2} - 3.925T_{ew} - 0.2302T_c + 179.51 \qquad (6.34)$$

The DATA/FORECAST/What-If-Analysis/Data Table command is used along with Equation 6.34 to construct a table of results for the combined regression analysis, and is shown in Figure 6.34.

It should be noted that a certain amount of art is involved in effecting a combined regression analysis. In most cases, the starting point is a construction of graphs of the result y as functions of each independent variable. The appearance of the curves obtained will suggest the functional form of the correlation to be tried.

Example 6.16: Comparison of Results of Examples 6.13 through 6.15

Examples 6.13 through 6.15 presented three methods for obtaining a regression analysis or correlation fit for the data shown in Figure 6.26. In this section, we will compare

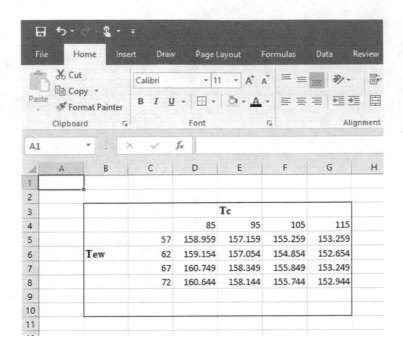

FIGURE 6.32

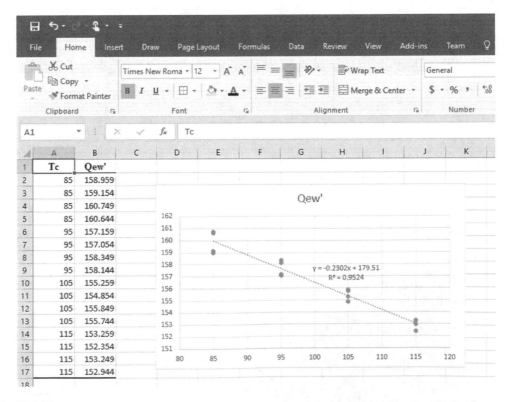

FIGURE 6.33

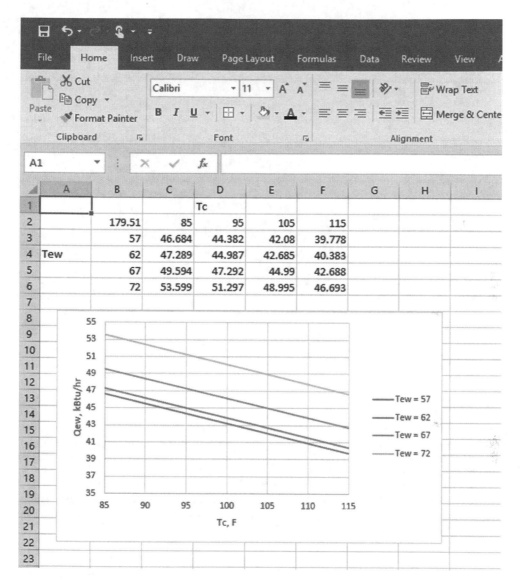

FIGURE 6.34

the regression results on the basis of a calculated root-mean-square deviation from the original data. Thus, we will perform the following calculation for each of the three regression results:

$$\text{RMS Dev } Q_{ew}(\text{regression}) = \left\{ \left[\Sigma (Q_{ew,regress} - Q_{ew,data})^2 \right] / n \right\}^{1/2} \qquad (6.35)$$

where the summation is taken over the 16 original data points of Figure 6.26 and $n=16$. The worksheet for the calculation is set up initially as shown in Figure 6.35, with the original data entered in columns A, B, and C and rows 5 through 20. The formula for the linear regression of Q_{ew} was given in Equation 6.28 and is entered in the worksheet in cell D5. The result for the exponential regression given in Equation 6.31

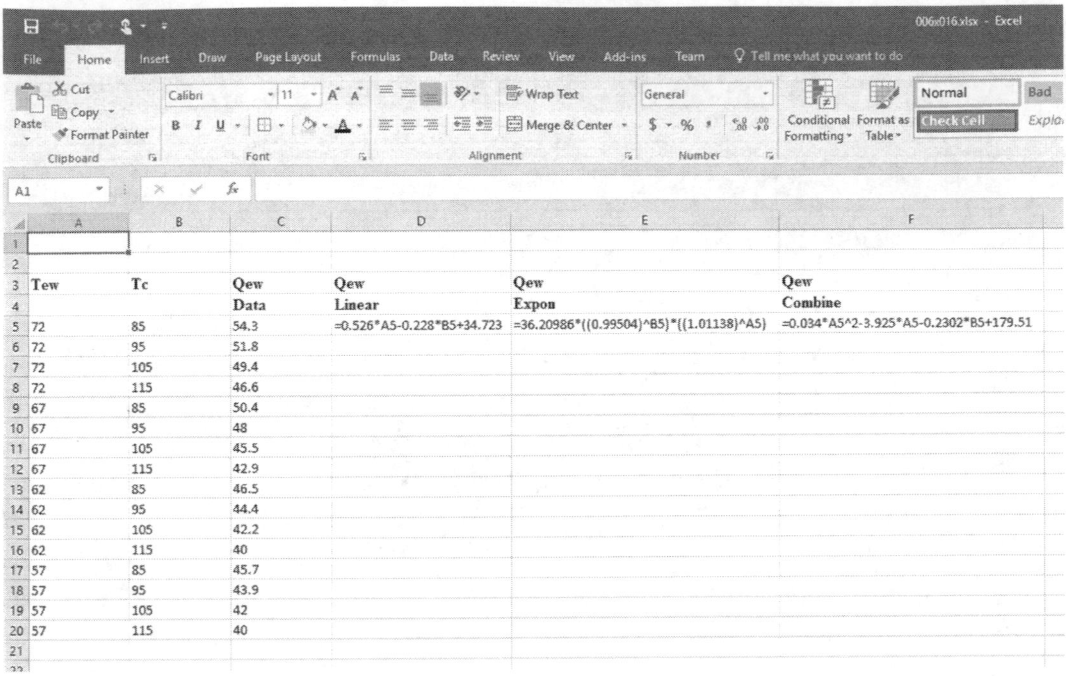

	A	B	C	D	E	F
1						
2						
3	Tew	Tc	Qew	Qew	Qew	Qew
4			Data	Linear	Expon	Combine
5	72	85	54.3	=0.526*A5-0.228*B5+34.723	=36.20986*((0.99504)^B5)*((1.01138)^A5)	=0.034*A5^2-3.925*A5-0.2302*B5+179.51
6	72	95	51.8			
7	72	105	49.4			
8	72	115	46.6			
9	67	85	50.4			
10	67	95	48			
11	67	105	45.5			
12	67	115	42.9			
13	62	85	46.5			
14	62	95	44.4			
15	62	105	42.2			
16	62	115	40			
17	57	85	45.7			
18	57	95	43.9			
19	57	105	42			
20	57	115	40			
21						

FIGURE 6.35

is entered in cell E5, and the formula for the combined regression, Equation 6.34, is entered in cell F5. In these formulas, A5 and B5 refer to the T_{ew} and T_c data values, respectively.

The formulas in D5:F5 are dragged-copied for another 15 rows to D20:F20. The display is switched from formula view to value view (via the CTRL+` key sequence), and the calculated values for the three $Q_{ew,regress}$ are displayed in D5:F20, as shown in Figure 6.36. In addition, the squares of $Q_{ew,regress} - Q_{ew,data}$ are computed and displayed in columns G, H, and I of this figure.

Finally, Equation 6.35 for the linear, exponential, and combined regressions is entered in cells N5, N7, and N9, respectively, as shown in Figure 6.37. In these formulas, the SUM worksheet function sums the squares of the deviations computed in columns G, H, and I. When the formulas are removed from view, the results are displayed in cells N5, N7, and N9 shown in Figure 6.38 and are repeated as

$$\text{RMS Dev } Q_{ew}(\text{linear}) = 0.9603 \tag{6.36}$$

$$\text{RMS Dev } Q_{ew}(\text{exponential}) = 0.8633 \tag{6.37}$$

$$\text{RMS Dev } Q_{ew}(\text{combined}) = 0.5578 \tag{6.38}$$

From these results, we can conclude that the combination regression procedure of Example 6.15 provides the best fit to the original data. Note that the combination regression is more complicated and requires more art and insight on the part of the person performing the analysis.

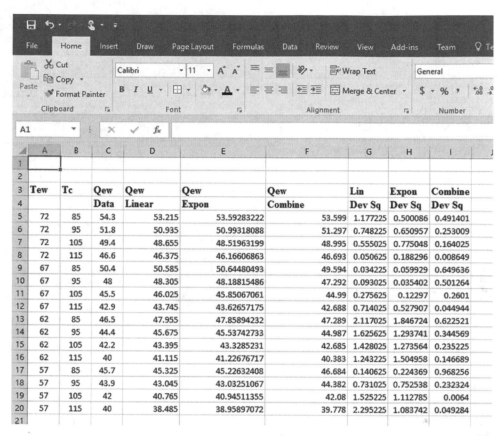

FIGURE 6.36

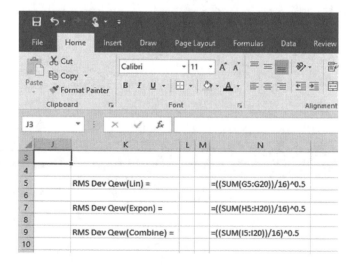

FIGURE 6.37

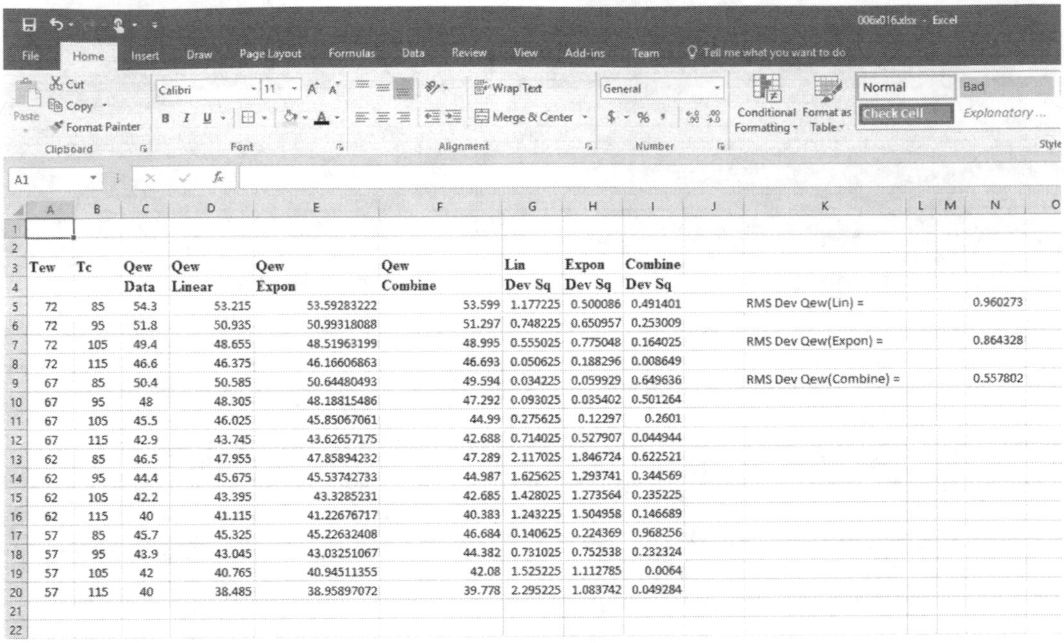

Tew	Tc	Qew Data	Qew Linear	Qew Expon	Qew Combine	Lin Dev Sq	Expon Dev Sq	Combine Dev Sq		
72	85	54.3	53.215	53.59283222	53.599	1.177225	0.500086	0.491401	RMS Dev Qew(Lin) =	0.960273
72	95	51.8	50.935	50.99318088	51.297	0.748225	0.650957	0.253009		
72	105	49.4	48.655	48.51963199	48.995	0.555025	0.775048	0.164025	RMS Dev Qew(Expon) =	0.864328
72	115	46.6	46.375	46.16606863	46.693	0.050625	0.188296	0.008649		
67	85	50.4	50.585	50.64480493	49.594	0.034225	0.059929	0.649636	RMS Dev Qew(Combine) =	0.557802
67	95	48	48.305	48.18815486	47.292	0.093025	0.035402	0.501264		
67	105	45.5	46.025	45.85067061	44.99	0.275625	0.12297	0.2601		
67	115	42.9	43.745	43.62657175	42.688	0.714025	0.527907	0.044944		
62	85	46.5	47.955	47.85894232	47.289	2.117025	1.846724	0.622521		
62	95	44.4	45.675	45.53742733	44.987	1.625625	1.293741	0.344569		
62	105	42.2	43.395	43.3285231	42.685	1.428025	1.273564	0.235225		
62	115	40	41.115	41.22676717	40.383	1.243225	1.504958	0.146689		
57	85	45.7	45.325	45.22632408	46.684	0.140625	0.224369	0.968256		
57	95	43.9	43.045	43.03251067	44.382	0.731025	0.752538	0.232324		
57	105	42	40.765	40.94511355	42.08	1.525225	1.112785	0.0064		
57	115	40	38.485	38.95897072	39.778	2.295225	1.083742	0.049284		

FIGURE 6.38

Problems

6.1 Planck's blackbody radiation formula is

$$E_{b\lambda} = C_{1\lambda}^{-5}/[\exp(C_2/\lambda T) - 1]$$

where

$C_1 = 3.743 \times 10^8$ and $C_2 = 14387$ and λ is in μm and T in K

Perform a numerical integration of this formula from 0 to 10 μm for temperatures of 1000 and 2000 K, i.e., evaluate $\int E_b \lambda d\lambda$ over the specified range. Repeat for $0 < \lambda < 20$ μm.

6.2 A set of experimental data is expected to follow the relationship

$$y = Cx^n$$

with the values of C and n given in the following table. Using appropriate IF statements, devise a worksheet to calculate the values of y over the range of 4 < x < 400,000. Then, plot y vs. x on a log–log scatter graph.

X	C	n
4–40	0.911	0.385
40–4,000	0.683	0.466
4,000–40,000	0.193	0.618
40,000–400,000	0.0266	0.805

6.3 Another set of experimental data is expected to follow the same functional relation as given in Problem 6.2. The values of C and n for these data are given in the following table. Devise a worksheet to calculate y as a function of x using IF statements. Then, plot the relation over the range $0.01 < x < 10^{12}$.

x	C	n
0.01–100	1.02	0.148
100–10,000	0.85	0.188
10,000–1.0E7	0.48	0.25
1.0E7–1.0E12	0.125	0.3333

6.4 On a scale of 0–100, student test scores have the values:

87, 73, 90, 75, 96, 76, 84, 57, 58, 20, 76, 56, 82, 95, 94, 52, 82, 89, 56, 77, 20, 98, 68, 98, 38, 47, 93, 59, 69, 68, 49, 80

Following the procedure of Example 6.5, arrange the scores in 10-point bins from scores ranging from 49.9 to 99.9, and construct an appropriate histogram of the scores. Determine the mean, median, maximum, and minimum scores as well as the sample and population standard deviations. Using histograms and variable bin widths, devise a system for setting letter grades of A through F for these scores. Consider at least three alternatives.

6.5 Compare a cumulative frequency distribution of the student scores in Problem 6.4 with that of a normal distribution having the same mean and standard deviation values. What do you conclude?

6.6 A certain experimental result y is a function of three measured variables through the relation:

$$y = x_1 x_2 / x_3$$

The measured variables have uncertainties of $w_{x1} = \pm 1.0$, $w_{x2} = \pm 0.5$, and $w_{x3} = \pm 0.2$, and the variables are measured over the following ranges:

$$5 < x_1 < 100$$

$$5 < x_2 < 100$$

$$15 < x_3 < 100$$

Devise a worksheet to calculate the values of y and their percentage uncertainties. Then, perform these calculations at the upper and lower limits of the measurement ranges and two intermediate points.

6.7 An experimental result y has the functional form of

$$y = x_1 + (x_2 x_3)^{1/2}$$

in terms of the three measured variables. The uncertainties in all three variables are ± 1.0, and the variables are measured over ranges of 10–100. Devise a worksheet to calculate values of y and their percentage uncertainty. Perform the calculations for the lower and upper limits of the measured variables and two intermediate points.

6.8 For the student test scores in Problem 6.4, plot the normal probability density function for one and two standard deviations.

6.9 The manufacturer's performance data for an air-conditioning unit similar to that discussed in Example 6.13 are given in the following table. Note that the electric work input W, expressed in kW, is furnished in addition to the cooling performance Q_{ew} in kBtu/h. Following the methods of Examples 6.13 and 6.14, obtain linear and exponential regressions for both Q_{ew} and W. Plot the results of the analysis compared with the data table. Compare the results of the regression analyses using RMS Dev defined in Equation 6.32.

Performance of an Air-Conditioning Unit			
T_{ew} (°F)	T_c (°F)	Q_{ew} (kBtu/h)	W (kW)
72	85	66.2	4.98
72	95	63.3	5.47
72	105	60.2	6.06
72	115	57	6.69
67	85	61	4.87
67	95	58.4	5.4
67	105	55.5	5.98
67	115	52.5	6.6
62	85	56.2	4.8
62	95	53.8	5.33
62	105	51.2	5.9
62	115	48.5	6.52
57	85	54.2	4.78
57	95	52.2	5.31
57	105	50.1	5.89
57	115	47.9	6.51

7

Financial Functions and Calculations

7.1 Introduction

In engineering design applications, different economic alternatives must frequently be investigated. Excel offers over 50 built-in financial functions that may be employed for such investigations. We shall now give examples of calculations based on some of these functions. First, we define the nomenclature employed in the syntax statements of the Excel functions and see how they relate to some elementary compound interest formulas.

7.2 Nomenclature

fv   This is the future value an investment may have after all payments have been made or accrued, including an initial payment.

nper   This is the number of periods that investment payments are made. The periods may be uniform or nonuniform in time and may be specified in terms of days, months, years, or the time between specific dates. This term is often designated by the symbol n.

pmt   This is the periodic payment to or from an investment medium. The payment may be constant or variable with time.

pv   This is the present value of a set of future payments or the current value of a loan.

Rate   This is the interest rate or discount rate of a loan or investment specified for a particular period, such as annually or monthly. It is usually designated by the symbol I. When entering the interest or discount rate as an argument in Excel functions, it must be expressed as a percentage or a decimal fraction. Twelve percent would be entered as 12% or 0.12, but not as 12. If 12% is stated as the annual rate, the monthly rate would be 12%/12 = 1% unless monthly compounding is specified.

Type   This is the type of interval for which investments are made, i.e., whether at the beginning or end of the period. Set the type as 0 (or omit the parameter) for payments at the end of the period or set the type as 1 for payments at the beginning of the period.

Dates   This is the entry of a specific date such as 4/15/2017 for April 15, 2017. Press F1 for Help within Excel, and then search for the term "date format" for a full discussion of Excel date formatting.

7.3 Compound Interest Formulas

The Excel financial functions are related to elementary compound interest formulas:

1. Amount to which $1 will accumulate at interest rate I per period for n periods:

$$fv = (1+I)^n \qquad (7.1)$$

2. Payment per period at interest rate I per period for n periods to repay a loan having a present value of $1:

$$pmt = I / [1 - (1+I)^{-n}] \qquad (7.2)$$

3. Present value of $1 contributed per period for n periods at interest rate I per period:

$$pv = [1 - (1+I)^{-n}] / I \qquad (7.3)$$

4. Amount to which $1 per period will accumulate when invested for n periods at interest rate I per period:

$$fv = [(1+I)^n - 1] / I \qquad (7.4)$$

All of these equations may be evaluated easily with a calculator for single-use computations, but the appropriate Excel function will be preferable for use in larger programs. In most cases, the Excel functions allow for more variables and flexibility than the simple formulas.

Table 7.1 gives a summary of the functions we will discuss, the syntax and parameters for evaluation of each function, and a description of the computation performed by the function. Parameters enclosed in square brackets ([]) are considered optional. At least one specific example will be given for each function, and, where needed, a copy of the appropriate worksheet displays will be provided.

Example 7.1: Accumulation of Uniform Monthly Payments

To what future value will a uniform series of monthly payments of $250 accumulate over a 3-year period with an annual interest rate of 9%?

The calculation can be made with either Equation 7.4 or the FV function.

The number of monthly periods is $3 \times 12 = 36$ and the monthly interest rate is $9\%/12 = 0.75\%$. Using Equation 7.4 and multiplying by the $250 monthly payment gives

$$FV = 250 \times [(1+0.0075)^{36}] - 1 / 0.0075 = \$10,288.18 \qquad (7.5)$$

The FV function yields

$$= FV(0.0075, \ 36, \ 250) = \$10,288.18 \qquad (7.6)$$

where the present value is assumed to be zero.

TABLE 7.1

Financial Functions and Operational Syntax

Name of Function	Syntax	Calculation Performed by Function
Financial Functions		
Present value	PV(rate, nper, pmt, [fv], [type])	Determines the present value of a set of uniform payments paid out over the number of periods, with a cash value of fv at the end of the last payment.
Future value	FV(rate, nper, pmt, [pv], [type])	Determines the future value of a set of uniform payments paid out of the investment medium (payments into the investment are negative values). The payments start with a lump-sum payment of pv.
Number of periods	NPER(rate, pmt, pv, [fv], [type])	Calculates the number of uniform payments needed to achieve the other values in the syntax statement.
Rate	RATE(nper, pmt, pv, [fv], [type], [guess])	Requires an iterative calculation with an initial guess for the final rate. If the guess is omitted, a value of 0.1 (10%) is assumed. The function determines the interest rate necessary to yield the other values specified in the syntax statement.
Payment	PMT(rate, nper, pv, [fv], [type])	Determines the constant periodic payment required to achieve the other values in the syntax statement.
Net present value	NPV(rate, value1, [value2], [... to 254 values])	Determines the sum of present values for up to 254 values at equally spaced time increments. The interest rate is assumed constant for each period. Payments received are entered as positive values, whereas payments made are entered as negative values. The values are entered at the end of each period. An initial investment at the start of the first period is added to the present value calculated by this function. This function may be thought of as the inverse of the IRR function, i.e., NPV(IRR(...), ...)=0.
Internal rate of return	IRR(values, [guess])	Requires an iterative calculation and an initial guess for the rate. If no guess is entered, a value of 0.1 (10%) is assumed. If #NUM! Error values appear, or an unacceptable result is obtained, repeat using a different guess. This function determines the rate of return per period necessary to return the set of cell values listed in the syntax. Payments received are entered as positive values, whereas payments to the investment are represented as negative values. The payments need not be uniform, but the period must be constant in time length. For the calculation, at least one positive and one negative value are required. This function may be thought of as the inverse of the NPV function, i.e., IRR(NPV(...), ...)=0.
Net present value of a series of payments that may be nonuniform	XNPV(rate, values, dates)	Determines the present values for a series of future cash flows that may be nonperiodic. Thus, it is necessary to specify the date when each cash payment is received. The first date must correspond to the first cash payment. All other data must have dates later than this entry. The number of values must equal the number of dates; otherwise, a #NUM! Error will be returned.
Modified internal rate of return	MIRR(values, finance_rate, reinvestment_rate)	The values argument represent a series of periodic payments (negative values) or income (positive values). The finance_rate is the "cost" of payments into the fund, whereas reinvestment_rate represents the rate at which payments out of the fund can be reinvested. The values must contain at least one positive and one negative value. The values may be nonuniform, but the period must be uniform.

(Continued)

TABLE 7.1 (*Continued*)

Financial Functions and Operational Syntax

Name of Function	Syntax	Calculation Performed by Function
Internal rate of return for nonperiodic payments	XIRR(values, dates, [guess])	Similar to IRR except that the period may be nonuniform. The IRR function should be used for periodic payments. At least one positive cash flow (payment out) and one negative cash flow (payment in) are necessary for the iterative calculation. If no guess is made for the interest rate, a value of 0.1 will be assumed. Each date entry must correspond to its respective value entry. The earliest date must be the first entry, but later dates can be in any order. The XIRR function is related to the XNPV function. The rate calculated by XIRR corresponds to an interest rate that will cause XNPV=0.

Example 7.2: Accumulation of Uniform Payments with Present Cash Value

Repeat Example 7.1 for a present cash value of $5000 that accumulates along with the monthly cash payments.

The FV function in this case becomes

$$= FV(0.0075,\ 36,\ 250,\ 5,000) = \$16,831.41 \tag{7.7}$$

Equation 7.1 may be used to calculate the future value for the $5000 as

$$FV = 5000 \times (1+0.0075)^{36} = \$6543.23 \tag{7.8}$$

When this value is added to that of Equation 7.5 of Example 7.1, we obtain $6543.23+ $10288.18=$16831.41, a value in agreement with the calculation using the FV function.

Example 7.3: Time Period for Defined Accumulation

A person wishes to accumulate a sum of $200,000 starting with an initial investment of $5000 and then making monthly payments of $1500. She assumes an interest rate of 12% per year. How many months will be needed to accumulate $200,000?

In this case, we apply the NPER function with the following arguments:

Rate = 9%/12 = 0.75%, pmt = $1500 per month, pv = $5000, fv = $200,000, and type = 0 (indicating payment at the end of the periods). Thus,

$$= NPER(0.75\%,\ -1500,\ -5000,\ 200,000,\ 0) = 89.46\ \text{months} \tag{7.9}$$

Note that we could also have taken rate=0.0075.

Example 7.4: Rate of Return for Discount Bonds

An investor buys a $250,000 bond with a coupon rate of 3% paid semiannually. The bond is purchased at a 20% discount ($200,000 for a $250,000 bond), and the bond matures in 7 years. What is the internal rate of return for this investment?

This example is an application of the IRR function. We first list all the payments paid out of the investment in column B of the worksheet in Figure 7.1. There are 14 semi-annual payments out (positive values) of $3750 ((3%/2)×$250,000) plus the $250,000 paid at maturity. In addition, there is a negative payment at the start of −$200,000,

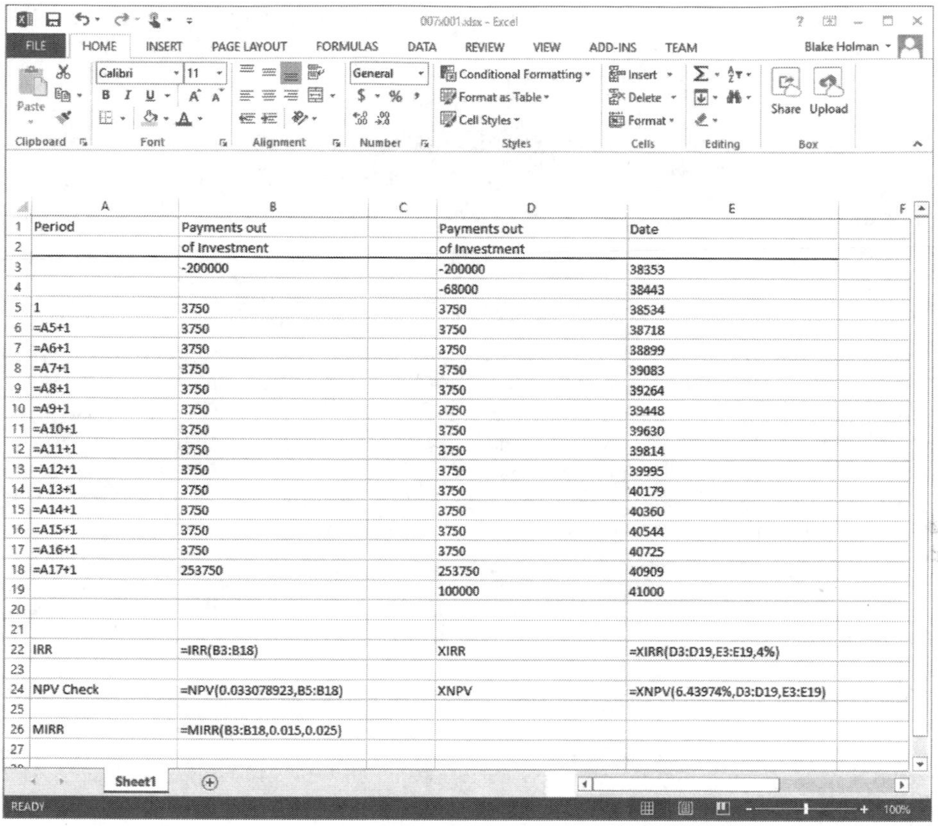

	A	B	C	D	E
1	Period	Payments out		Payments out	Date
2		of Investment		of Investment	
3		-200000		-200000	38353
4				-68000	38443
5	1	3750		3750	38534
6	=A5+1	3750		3750	38718
7	=A6+1	3750		3750	38899
8	=A7+1	3750		3750	39083
9	=A8+1	3750		3750	39264
10	=A9+1	3750		3750	39448
11	=A10+1	3750		3750	39630
12	=A11+1	3750		3750	39814
13	=A12+1	3750		3750	39995
14	=A13+1	3750		3750	40179
15	=A14+1	3750		3750	40360
16	=A15+1	3750		3750	40544
17	=A16+1	3750		3750	40725
18	=A17+1	253750		253750	40909
19				100000	41000
20					
21					
22	IRR	=IRR(B3:B18)		XIRR	=XIRR(D3:D19,E3:E19,4%)
23					
24	NPV Check	=NPV(0.033078923,B5:B18)		XNPV	=XNPV(6.43974%,D3:D19,E3:E19)
25					
26	MIRR	=MIRR(B3:B18,0.015,0.025)			
27					

FIGURE 7.1
Rate of return for discount bonds

which represents the purchase price of the bond. The last payment is therefore \$250,000 + \$3750 = \$253,750. The IRR function is entered in cell B22 as

$$= \text{IRR(B3 : B18)} = 3.3078923\% \tag{7.10}$$

This is the semiannual rate; thus, the annual rate would be 3.3078923×2=6.6157846%. Note that no guess was entered for the rate, so the function assumed a value of 10%. If the calculation were made with a guess of 2%, the same answer would result.

The answer may be checked by calculating the NPV function using a rate of 3.3078923% and using the same payments in the IRR calculation. This computation is also listed on the worksheet at cell B24, and it gives the result

$$= \text{NPV}(3.3078923\%,\ \text{B2 : B15}) = \$200,000.00 \tag{7.11}$$

which, of course, is the purchase price of the bond at the start of the first period.

Example 7.5: Rate of Return with Reinvestment of Bond Interest

Assume that the funds received by the bond investor in Example 7.4 may be reinvested at a rate of 5% per year. The interest rate for financing the original investment of \$200,000 is 3% per year. Calculate the modified rate of return for this financing arrangement.

The calculation is performed by calling up the MIRR function in cell B26 of Figure 7.1. For the semiannual period, the financing rate is entered as 0.015 (3%/2), whereas the reinvestment rate is entered as 0.025 (5%/2). We have the result

$$= \text{MIRR(B1 : B15, 0.015, 0.025)} = 3.2260364\% \qquad (7.12)$$

or an annual rate of $2 \times 3.2260364 = 6.4520728\%$.

Example 7.6: Rate of Return of Zero-Coupon Bond

The investor in Example 7.4 decides to buy another 7-year bond, 3 months after the first bond purchase. The new bond is a "zero-coupon" instrument, which sells at a deep discount but does not return any funds to the investor until maturity. The purchase price for this bond is $68,000 for a value at maturity of $100,000 5 years later. Calculate the internal annual rate of return for this bond.

For this calculation, we can use the compound interest formula of Equation 7.1 and solve for I, the interest rate. We have:

$$pv = 68,000$$
$$fv = 100,000$$
$$N = 5$$

Using the logarithms in Equation 7.1 results in an equation for the interest rate as

$$I = \exp[\ln(fv/pv)/n] - 1$$

$$I = \exp[\ln(100,000/68,000)/5] - 1 = 0.080185 = 8.0185\%$$

The 8.018% is an annual rate because n=5 is the number of annual periods.

Example 7.7: Rate of Return of Multiple Bond Investments

The bond purchases in Examples 7.4 and 7.6 may be considered as one investment with variable payouts over the period of 7 years 3 months. (The second bond is purchased 3 months after the first bond.) Calculate the internal rate of return for this overall investment.

This problem involves nonequal payments at nonuniform periods. We list the payments in column D of Figure 7.2, with their corresponding dates listed in column E. Based on a 365-day year, the computer converts the date for each payment to a number of days beginning with a zero day starting on January 1, 1900. The first bond purchase is listed as −$200,000 in cell D3, and the second bond purchase is listed as −$68,000 in cell D4. The coupon payments for the first investment are listed in cells D5:D17. The maturity payment for the first bond is the entry in cell D18, whereas the maturity payment for the second bond is listed as $100,000 in cell D19. The XIRR function is employed for the calculation with a guess of 4% and is entered in cell E22 as

$$= \text{XIRR(D3 : D15, E3 : E15, 4\%)} \qquad (7.13)$$

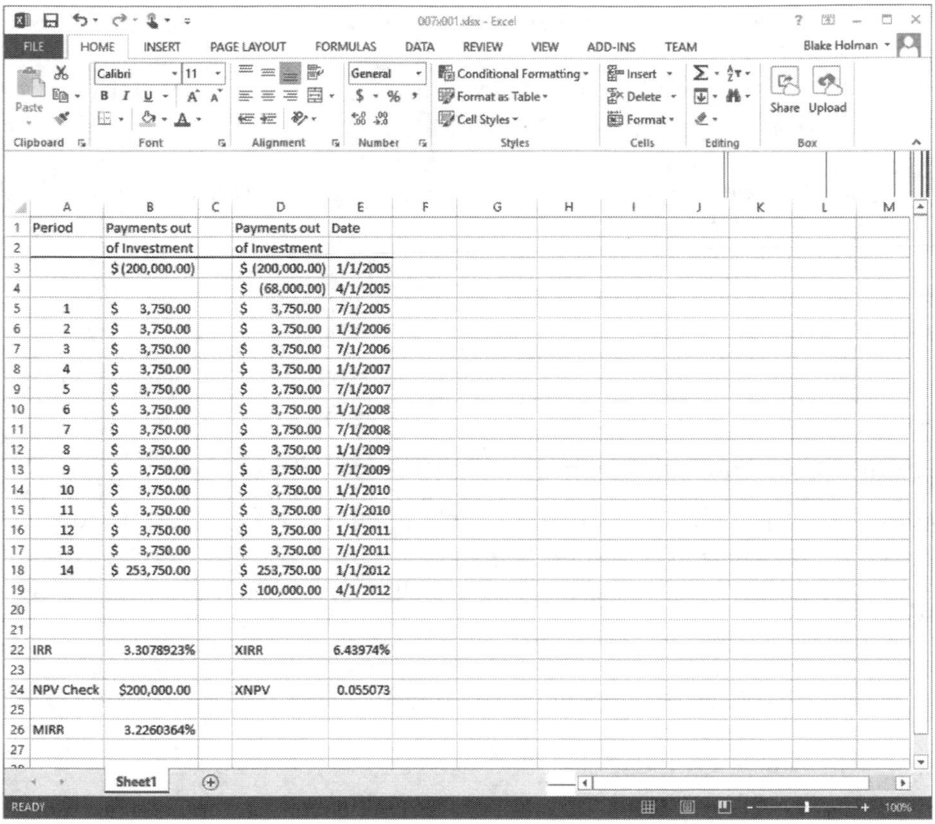

	A	B	C	D	E
1	Period	Payments out		Payments out	Date
2		of Investment		of Investment	
3		$ (200,000.00)		$ (200,000.00)	1/1/2005
4				$ (68,000.00)	4/1/2005
5	1	$ 3,750.00		$ 3,750.00	7/1/2005
6	2	$ 3,750.00		$ 3,750.00	1/1/2006
7	3	$ 3,750.00		$ 3,750.00	7/1/2006
8	4	$ 3,750.00		$ 3,750.00	1/1/2007
9	5	$ 3,750.00		$ 3,750.00	7/1/2007
10	6	$ 3,750.00		$ 3,750.00	1/1/2008
11	7	$ 3,750.00		$ 3,750.00	7/1/2008
12	8	$ 3,750.00		$ 3,750.00	1/1/2009
13	9	$ 3,750.00		$ 3,750.00	7/1/2009
14	10	$ 3,750.00		$ 3,750.00	1/1/2010
15	11	$ 3,750.00		$ 3,750.00	7/1/2010
16	12	$ 3,750.00		$ 3,750.00	1/1/2011
17	13	$ 3,750.00		$ 3,750.00	7/1/2011
18	14	$ 253,750.00		$ 253,750.00	1/1/2012
19				$ 100,000.00	4/1/2012
20					
21					
22	IRR	3.30789230%		XIRR	6.43974%
23					
24	NPV Check	$200,000.00		XNPV	0.055073
25					
26	MIRR	3.2260364%			
27					

FIGURE 7.2
Rate of return of multiple bond investments

with a result of 6.43974% on an annual basis. The 4% entry is the guess to start the iterative calculation. The result may be checked with the XNPV function entered in cell E24 as

$$= \text{XNPV}\ (9.3005\%,\ \text{D3:D15},\ \text{E3:E15}) = 0.055073 \tag{7.14}$$

which is nearly 0 (compared to 100,000). In other words, the present value of all the payments is zero until the initial investments are made.

7.4 Investment Accumulation with Increasing Annual Payments

The first case we will consider is the accumulation problem. The nomenclature for this case is as follows:

A Present accumulation at beginning of period
N Number of periods that payments are to be made into investment
S Amount to be invested at the beginning of initial period

R Rate of increase in S in subsequent years (R=0.05 for 5%) (rate may be either positive or negative)

I   Assumed rate of return (interest) on investment (I=0.05 for 5%) for the period

A(n) Accumulation of investment after n periods

By writing out the sequence of payments for n periods, it is possible to express the relation for A(n) in a compact formula as:

$$A(n) = A \times (1+I)^n + S \times (1+I) \times [(1+I)^n - (1+R)^n] / (I-R) \tag{7.15}$$

7.5 Payout at Variable Rates from an Initial Investment

The second case is just the reverse of the accumulation problem. In this instance, we start out with an investment sum and then make a payout that increases in each period for a specified number of periods. The nomenclature employed for this case is:

B Accumulation at start of payout

N Number of periods payout is to be made

P Amount of first payment at end of first period

C Rate of increase in P of each period (C=0.05 for 5%)

I   Assumed rate of return (interest) on investment for each period (I=0.05 for 5%)

B(n) Capital (funds) remaining at end of n years

The relation for B(n) is:

$$B(n) = B \times (1+I)^n - P \times [(1+I)^n - (1+C)^n] / (I-C) \tag{7.16}$$

There are two limiting cases of Equation 7.11 that express the fraction P/B for (a) no reduction in capital, i.e., B(n)=B and (b) complete consumption of capital, i.e., B(n)=0. For these cases, we have:

$$P/B = (1-C) \times [1 - (1+I)^{-n}] / \{1 - [(1+C)/(1+I)]^n\} \tag{7.17a}$$

for B(n)=B
  and

$$P/B = (1-C) / \{1 - [(1+C)/(1+I)]^n\} \tag{7.17b}$$

for B(n)=0
  A more general relation for the retention of a fraction F of the capital uses F=B(n)/B. This results in:

$$P/B = (1-C) \times [1 - F \times (1+I)^{-n}] / \{1 - [(1+C)/(1+I)]^n\} \tag{7.17c}$$

Example 7.8: Accumulation of Retirement Contributions with Increasing Rate

The purpose of this example is to show the effect of increasing contributions to an investment each year at some assumed rate of increase. The results can be rather dramatic. Suppose that a person has accumulated $25,000 in savings toward retirement at age 35 and is now starting a contribution plan in which he will be able to contribute $5000 the first year. Because of anticipated salary increases, he estimates that he will be able to increase his contributions by 3% each year (R = 0.03). He estimates that he will be able to achieve an average return on investment of 7% over the years (I = 0.07). He wishes to estimate the accumulation that will be achieved at age 65 (n = 30 years) using Equation 7.15. We have already noted the values of R and I earlier. Observe that the present accumulation A is $25,000, whereas the initial annual contribution is S = $5000. Inserting all these values in Equation 7.15 gives the accumulation A(n) after 30 years:

$$A(n) = \$883,799.13$$

which is a rather substantial amount. This sum might be entered into Equation 7.16 with appropriate assumptions to estimate retirement benefits over an extended period of time.

Example 7.9: Payout of Retirement Benefit That Increases with Cost of Living

In this example, we start with an assumed accumulation and examine the effects of drawing down the sum with increasing payouts to match an anticipated rise in the cost of living. For this case, we assume that an investment is available that will yield a tax-free return of 5% (I = 0.05) after taxes and that the cost of living will increase 3% per year (C = 0.03). The prospective retiree has accumulated a sum of $1,500,000 and has determined that she needs about $50,000 per year for a comfortable retirement. We use Equation 7.16 for the calculation, taking I and C as given along with B = $1,500,000 and P = $50,000. The prospective retiree wishes to know what will remain of her original investment after 20 years (n = 20). We insert the aforementioned values in Equation 7.16 and obtain:

$$B(n) = \$1,861,980.38$$

i.e., her investment has actually grown over this period of time by the amount of $361,980.38. Please note that this calculation assumes that the $1,500,000 initial investment has been accumulated after taxes.

Example 7.10: Payout to Exhaust Retirement Fund

The individual in Example 7.9 wishes to determine the maximum initial payment for the given conditions that will just exhaust the fund after 20 years. Equation 7.17b applies, with I=0.05, C=0.03, and n=20, to give

$$P/B = 0.062638$$

With B=$1,500,000, this gives a first-year payment of 0.062638×$1,500,000 or

$$P = \$93,957.00$$

A worksheet is shown in Figure 7.3 listing the input parameters and formulas necessary for the solution of the previous examples. To use the program, the formula display can be removed by clicking CTRL+`, which toggles viewing of formulas on and off within

FIGURE 7.3
Accumulation and payout of investments.

the worksheet. The values of the parameters are entered at B4:B8, B13:B17, and B20. The results are displayed at B10, B18, or B21. *Do not enter numerical values in B10, B18, or B21 as doing so will overwrite the formulas in those cells.*

Problems

7.1   Three $1500 monthly payments are made into an account with an initial balance of $15,000. Three withdrawals of $2000 per month and six deposits of $1600 per month follow these payments. Calculate the present and future values of these payments for an interest rate of 8% per year.

7.2 An investor plans to make 12 annual deposits of $6000 into an investment with a present value of $30,000. What rate of interest is required for the investment to accumulate to $250,000 in 12 years?

7.3 Using the results of Problem 7.2, apply the PMT function to verify the $6000 annual payment.

7.4 A bond that matures in 5 years has a present market value of $10,000 and a maturity value of $12,000. Semiannual coupon payments are made at an annual rate of 5% of the maturity value. Assuming the cash payments may be reinvested at a rate of 5%, calculate the present value of the investment and the value at maturity.

7.5 An investment group has a fund with a present cash value of $75,000. Monthly payments are to be added to the fund in the amount of $6000 per month. The anticipated rate of return on the investments is 8% per year. How long will it take to accumulate $350,000?

7.6 An investment group provides a cash outlay of $500,000 by borrowing the sum at an annual interest rate of 4.5% with monthly payback. The funds are invested in a real-estate venture with an anticipated return of 11% per year in semiannual payments. Calculate the accumulated value at the end of 8 years and the present value of the investment returns. State any assumptions.

7.7 A housing project requires an initial cash outlay of $600,000 followed by monthly payments of $30,000 for 24 months. The project will provide quarterly income payments (8 payments) of $200,000 each. What is the internal rate of return of this investment?

7.8 Financial advisors sometimes recommend "laddering" of fixed income investments. The following three investments are suggested:

- $135,000 in a bond maturing to $200,000 in 5 years with an annual payout of 6% of the maturity value

- $50,000 1 year later in another 5-year bond yielding 5.75% and selling for 80% of maturity value

- $65,000 a year later in a 5-year bond yielding 6.25% and selling for 87% of maturity value
    - Calculate a net present value of all the payments assuming a rate of 5%.
    - Calculate the internal rate of return of the investments over the 8-year period.

7.9 If the financing cost of the bond investments in Problem 7.8 is 5.25% and the reinvestment rate is 6%, calculate the modified rate of return.

7.10 A construction project is to be financed with a cash payment of $200,000 and a bank loan of $1,500,000 at 7%, paid back in 24 equal monthly installments. Income from the project is assumed to start at month 25 at the rate of $80,000 per month for 12 months followed by income of $120,000 per month for 24 months. Calculate the internal rate of return of all the payments over the 5-year period.

7.11 Assuming the payments from the project in Problem 7.10 are invested at a rate of 5.75%, what will be the accumulation at the end of the 5-year period? What present value, invested at 5.75%, would produce the same accumulation?

7.12 Compare the following investments on an appropriate basis. State any assumptions.

    a. A tax-free zero-coupon bond yielding 4% to maturity in 10 years.

    b. A tax-free coupon bond yielding 4% paid semiannually over a 10-year period.

    c. The same present value as (a) invested in a tax-sheltered account that accumulates at 6.15% for 10 years. The gain over the initial investment is taxable at a rate of 33% at the end of the 10-year period.

    d. The same present value as (a) invested in a taxable bond yielding 5.75%. The income is taxed at a rate of 33%.

    e. The same present value as (a) invested in a real estate venture, which accumulates at the rate of 7%, with the profit (gain) taxed at a rate of 20% at the end of 10 years.

7.13 A sum of $500,000 is available for payout with an investment return of 5% with the payout increased 2% each year over a period of 15 years. What initial payment is allowed if the fund is to be exhausted after 15 years?

7.14 A trust fund is to be created, which will accumulate to $500,000 over a 15-year period assuming a return on investment of 5%. Annual payments will be made into the fund, which increase at the rate of 2.5% per year. What initial payment is required? For a 5% interest rate, what is the present value of all payments?

7.15 A dedicated fund of $500,000 is provided for maintenance of a facility over a 20-year period. The facility will be torn down at the end of 20 years. It is estimated that the maintenance costs will increase at the rate of 3.1% per year and the funds will earn a rate of return of 5.7%. Based on these assumptions, what initial annual payment should be made to exhaust the fund? What will be the last payment?

7.16 A project is to be financed over a 20-year period. The assumption is made that income will rise each year such that the net cash input required will decrease by 3% per year. A 6% return on investment is assumed. The initial annual payment is $150,000 from an investment fund of $750,000. What will be the value of the fund after 20 years?

7.17 Suppose the investors for the project in Problem 7.16 desire to retain all of the initial capital of $750,000 after 20 years. What first-year payment would be permitted? Repeat for retention of 50% of the original capital.

7.18 A trust fund has an initial cash value of $250,000 available for investment. Additional annual payments into the fund will be made starting with $25,000. The annual payments will increase by 4% per year and the return on investment is assumed to be 6% per year. Compare two alternatives:

    a. The investments are made into a tax shelter in which the total interest gain is taxed at 33% after 10 years.

    b. The interest gain is taxed annually at a rate of 33% that reduces the effective after-tax return on investment to $(1 - 0.33) \times 6\% = 4.02\%$.

8

Optimization Problems

8.1 Introduction

We have seen how Solver and Goal Seek may be employed to solve single nonlinear equations or simultaneous linear or nonlinear equations through an iteration process. That process iterates on the values of the variables to cause a target function to approach zero. Now, we will see how Solver can be used to maximize or minimize a function subject to a set of constraints that take the form of inequalities. The function that is set as the target is usually called the *objective function*. Other titles for the objective function may be used depending on the situation they represent, such as cost function (minimize costs), weight function (minimize weight), profit function (maximize profits), etc. The optimization process is normally treated as a subject in operations research, and the mathematical techniques are part of the subject of mathematical programming.

Specific applications of the optimization processes are beyond the scope of this text; however, we direct the reader to References 1, 4, and 5 for additional information.

The objective function is defined as

$$y = g(x_1, x_2, \ldots, x_n) \tag{8.1}$$

which is expressed as a function of the x_i variables. The constraints on the g function are also expressed in terms of the x_i variables through additional functional relations, such as

$$\partial_i(x_1, x_2, \ldots, x_n) = 0, \leq 0, \text{ or } \leq 0 \tag{8.2}$$

The number of ∂_i functions may be small or large, and are generally not equal to the number of x_i variables. Note that both equalities and inequalities may be part of the constraints. For simple problems, the number of variables may number only two or three; for complicated problems, they may number in the hundreds or thousands. For only two variables, i.e., x_1 and x_2, the optimization process may be depicted graphically; for larger numbers of variables, such a representation is not possible (unless a person has extraordinary visualization capabilities!).

If both the g and ∂ functions are linear in x_i, we say that the problem is linear, and the solution is obtained by linear programming. If one or more of the g or ∂ functions is nonlinear, the problem is nonlinear and becomes more complicated. We shall consider two optimization problems, each with two variables (one linear and one nonlinear). As mentioned earlier, such problems may be represented graphically. We will solve these problems graphically to illustrate the concepts. Then, we will employ Solver to obtain the same solutions, independent of any graphical display.

8.2 Graphical Examples of Linear and Nonlinear Optimization Problems

We first consider the linear problem:
  Maximize or minimize

$$y = g = 2x_2 + 3x_1 \tag{8.3}$$

subject to the constraints

$$\partial_1 = x_1 + x_2 \geq 5$$

$$\partial_2 = 2x_1 + x_2 \leq 12 \tag{8.4}$$

$$\partial_3 = x_1 \leq 0$$

$$\partial_4 = x_2 \leq 0$$

In most practical problems, the variables x_i are positive because they represent cost per item, number of bolts or nuts produced, quantity of an ingredient in a food process, etc.

The graphical display for this problem is shown in Figure 8.1. First, the line $x_1 + x_2 = 5$ is plotted as shown. The acceptable values of y must lie above this line because of the ≤ 5 specification. Next, the line $2x_1 + x_2 = 12$ is plotted as indicated in the figure. The acceptable values of y must lie below this line because of the ≤ 12 specification on the function. Finally, the lines representing $x_1 = 0$ and $x_2 = 0$ are the coordinate axes of the figure, and the ≤ 0 specification indicates that only positive values are acceptable. As a result of these four constraints, we find that the acceptable values of y must lie within the four-sided figure indicated by the shaded lines. This area is called the *feasible region*.

Straight lines are then plotted on the chart for different values of $y = \text{constant} = 2x_2 + 3x_1$. These lines all have a slope of -1.5 and a y-intercept of $y/2$. All points on the line $y = 30$ lie

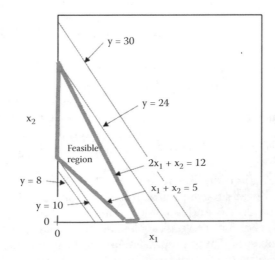

FIGURE 8.1

outside the feasible region, so $y=30$ is an unacceptable solution. The line $y=24$ has one point that lies just in the feasible region and intersects this region at the uppermost corner. There is no larger value of y that will lie in the feasible region, so $y=24$ is the maximum value for the stated constraints.

The line $y=8$ lies entirely below the feasible region and thus, cannot represent an acceptable solution. The line $y=10$ just intersects the feasible region at $x_1=0$ and $x_2=5$. There is no smaller value of y that will ensure the line has a point inside the feasible region; so, $y=10$ is the minimum value that will satisfy the given constraints.

It can be shown that for linear problems, acceptable minimum or maximum values of the objective function must always correspond to one of the vertices of the feasible region (see Reference 4). We can, therefore, state our final answers as:

$$y = y_{max} = 24 \text{ at } x_1 = 0 \text{ and } x_2 = 12$$

$$y = y_{min} = 10 \text{ at } x_1 = 0 \text{ and } x_2 = 5$$

Later, we will see how this problem can be worked using Solver.

The second optimization problem we consider has an objective function of

$$y = g = x_1 + x_2 \qquad (8.5)$$

and the constraints

$$\partial_1 = 2x_2 + 3x_1 \leq 30$$

$$\partial_2 = x_1^2 + x_2^2 \geq 36$$

$$\partial_3 = x_2 \leq 12 \qquad (8.6)$$

$$\partial_4 = x_1 \geq 0$$

$$\partial_5 = x_2 \geq 0$$

Because of the nature of ∂_2, this is a nonlinear problem. The graphical display for this problem is shown in Figure 8.2. First, the line $2x_2+3x_1=30$ is plotted as shown. Acceptable solutions must lie below this line because of the ≤ 30 constraint. The nonlinear (circle) constraint is plotted as a quarter circle, and the solution must lie to the top and right because of the $\partial_2 \leq 36$ specification. The line $x_2=12$ is drawn at the top of the figure, and acceptable solutions must lie below this line. Finally, the coordinate axes lines $x_1=0$ and $x_2=0$ are drawn, and the acceptable solutions are restricted to positive values. The feasible region is the five-sided figure enclosed by the shaded lines.

Values of the objective function $y=g=x_1+x_2$ for $y=$constant are now drawn as straight lines with a slope of -1 and a y-intercept of y. The line $y=15$ lies entirely outside the feasible region; so, $y=15$ is an unacceptable solution. The line $y=14$ just intersects the feasible region at the upper-right corner located at $x_1=2$ and $x_2=12$. This is the largest value of y that will lie in the feasible region. For smaller values of y, we note that the line $y=6$ just intersects the feasible region at two points: $x_1=6$, $x_2=0$ and $x_1=0$, $x_2=6$. Any smaller value

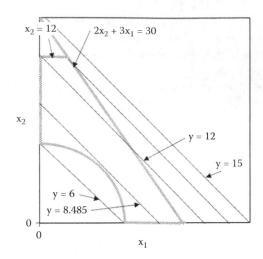

FIGURE 8.2

of y will plot the line outside the feasible region. Thus, we have two minimums. Our final results are as follows:

$$y = y_{max} = 14 \text{ at } x_1 = 2 \text{ and } x_2 = 12$$

$$y = y_{min} = 6 \text{ at } x_1 = 0, x_2 = 6 \text{ or } x_1 = 6, x_2 = 0$$

8.3 Solutions Using Solver

To obtain solutions to the example optimization problems, the Solver worksheet is organized in a fashion similar to that used for the solution of simultaneous equations. For the linear example, Figure 8.3a (displaying cell formulas) and Figure 8.3b (displaying cell

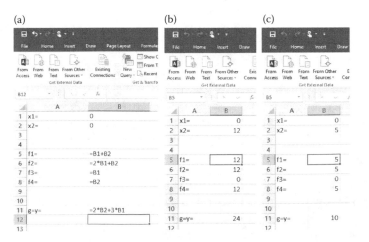

FIGURE 8.3

values) demonstrate the problem setup. The variables x_1 and x_2 are entered in cells B1 and B2. Solver will iterate on these values to obtain a solution. The constraint function formulas are entered in cells B5 through B8 without the inequality conditions. They will be added in the dialog box later. Finally, the objective (target) g function formula is entered in cell B11. At this point, the formula entries should be checked to ensure that they are correct.

Solver is now called up by clicking DATA/ANALYZE/Solver. The Solver parameters window appears as shown in Figure 8.4a. The target cell is set as the g function in cell B11, and the inequality constraints are entered in accordance with Equation 8.4. For the maximization part of the problem, "Max" is clicked in reference to the target function. The "By Changing Variable Cells" entries are the variable cells B1 and B2. Next, make sure that the

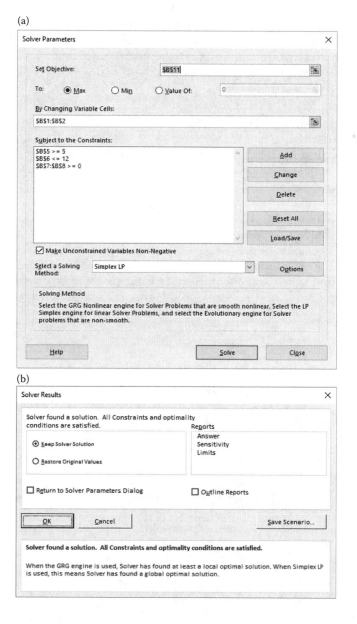

FIGURE 8.4

Simplex LP option is selected under the Select a Solving Method drop-down menu, as this problem is linear. Zero values are entered as the initial guesses for the variable cells B1 and B2. Click Solve and the Solver Results window will appear (Figure 8.4b). As well, the solution to the problem will appear in cells B1, B2, and B11 as shown in Figure 8.3b. The operation is repeated for the minimization condition. The solutions to the maximization problem may be taken as the initial guess for the minimization problem. The procedure is the same as that for the maximization problem except for clicking Min instead of Max. Again, the solutions will appear in cells B1, B2, and B11 as shown in Figure 8.3c. The answers are the same as those obtained graphically in Figure 8.1.

The solution for the nonlinear problem is obtained in the same way as that for the linear problem, except that the GRG Nonlinear option is selected in the Select a Solving Method drop-down menu. In this case, there are five constraint functions that are entered in cells B5 through B9. The formula for the objective g function is entered in cell B12.

The formulas for the maximization condition along with initial guesses of zero are shown in Figure 8.5a. The resultant solution for the maximization condition is shown in Figure 8.5b, and the values agree with the graphic solution obtained previously. The Solver Parameters window for the maximization condition, with entries of the respective constraints, is shown in Figure 8.6a. The two solutions for the minimization condition are shown in Figure 8.6b. The left-hand side solution is obtained with initial guesses of zero for cells B1 and B2. The right-hand (RH) side solution is obtained with guesses taken as the solutions to the maximization problem, i.e., $x_1=2$ and $x_2=12$. These results illustrate the fact that there may be more than one optimum solution when nonlinear constraints are involved. We see that the specific minimum point attained by Solver depends on the direction from which the solution is approached. For a real-life physical problem, the best answer may be selected on the basis of other factors.

In the nonlinear example just described, both minimum values of the objective function had the same value: $y=6$. Suppose the constraint $x_2 \le 0$ is changed to $x_2 \le 1.0$ and the initial guesses are $x_1=10$ and $x_2=6$. Now the minimum value of y will be displayed as $y=6.91608$ at the point $x_1=5.91608$ and $x_2=1.0$. Consulting Figure 8.2, we see that this is just the intersection of $x_2=1$ and $x_1^2+x_2^2=36$, i.e., $(36-1)^{1/2}=5.91608$.

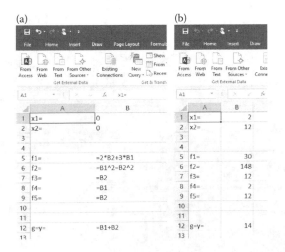

FIGURE 8.5

(a)

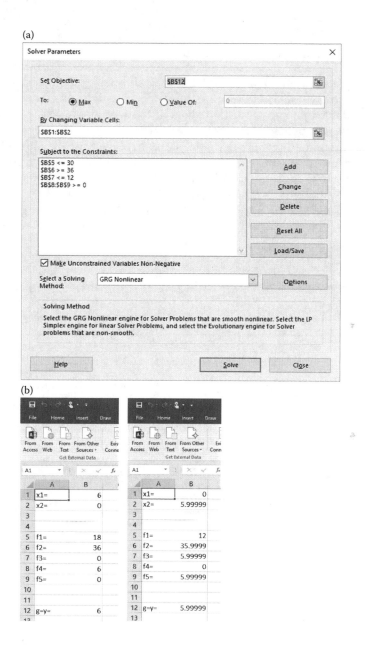

(b)

FIGURE 8.6

If the solution is approached using $x_1=0$ and $x_2=0$, the objective function is as before, which is 6.0. We say that the first point at $y=6.91608$ represents a local minimum, whereas the second point represents an absolute minimum. The minimum points depend on the way the values are approached in the mathematical process. For this problem, we have a graphical picture to guide us in assigning the names of local or absolute minimum values. The solution may not be so easy in problems with more than two variables. For nonlinear problems, the correct solution may hinge upon the physical insight brought to the analysis by an astute engineer.

8.4 Solver Answer Reports for Examples

Upon deciding that a satisfactory solution has been obtained, we click Save Scenario in the Solver Results dialog window. At the same time, we may call for the Answer Report, which will then be available as a separate titled sheet in the workbook. The Answer Reports for the linear optimization problem are shown in Figures 8.7 and 8.8. We note that the target (objective function) and adjustable cell values are displayed for the start of iteration (initial guesses), and the final values after all the constraints have been satisfied. The final ∂ function values are displayed in the Constraints box along with their respective constraint formulas. Under Status, the term *binding* means that the constraint applies to the final solution. The term *not binding* means there is a certain slack between the constraint and

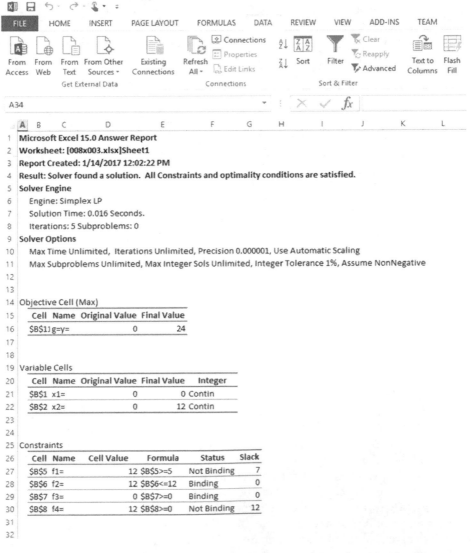

FIGURE 8.7

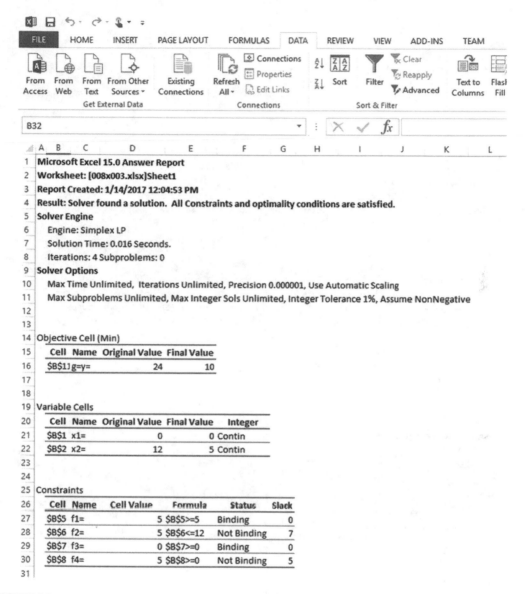

FIGURE 8.8

the final solution. These slack values are also displayed. In the maximization example (Figure 8.7), two of the constraints are binding and two are not binding. A zero slack results when a constraint is binding.

The Answer Report for the minimization example is shown in Figure 8.8. This report shows an initial value of the g function as 24—the final result of the maximization problem. The original values for the adjustable cells are also taken as the final values for the maximization problem. As in the maximization problem, two of the constraints are binding and two are not binding.

The Answer Reports for the nonlinear optimization problem are shown in Figures 8.9 and 8.10. The original guesses are taken as zero for the maximization problem (Figure 8.9),

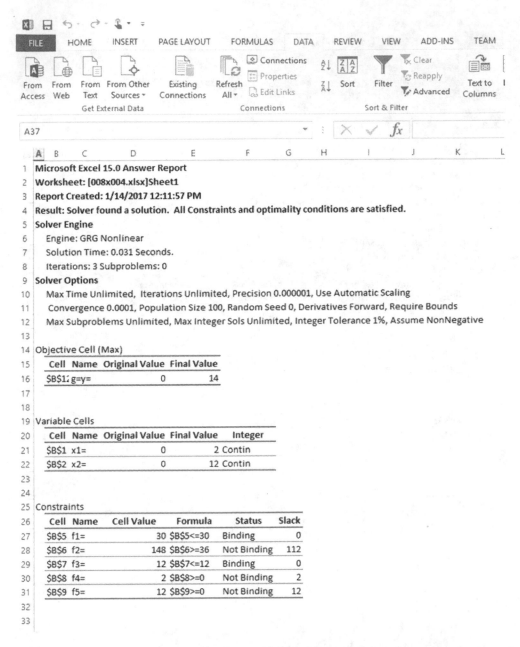

FIGURE 8.9

and the final values of the target and adjustable cell values are indicated. For this case, two of the constraints are binding and three are not binding. Again, when a constraint is binding, we note that its slack is zero. The minimization problem (Figure 8.10) starts with initial guesses taken as the answers from the maximization problem. There are two binding constraints for this case and three not-binding constraints.

The Answer Report sheets are preformatted in a style suitable for report presentation or transfer to another document.

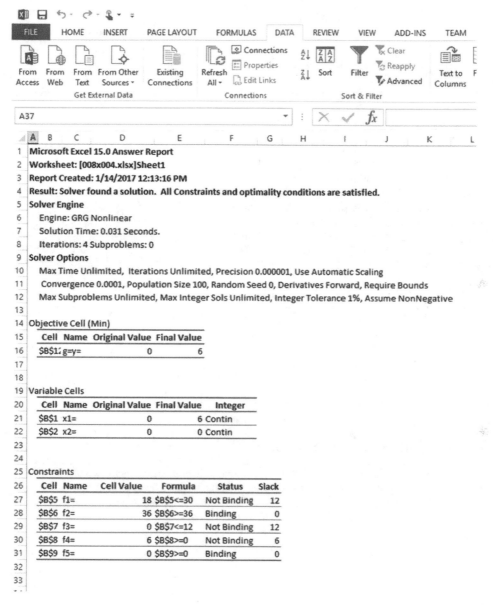

FIGURE 8.10

8.5 Nomenclature for Sensitivity Reports

Reduced gradient	This indicates the rate of change of the target cell per unit change in the changing cell.
Lagrange multiplier	This indicates the rate of change of the target cell per unit increase in the respective constraint.

If the Assume Linear Model box is checked, the following additional information is provided for each changing cell:

Reduced cost	This replaces the reduced gradient.
Objective coefficient	This indicates the relative relationship between a changing cell and the target cell.
Allowable increase	This indicates the change that must be present in the objective coefficient until there is a corresponding increase in the optimal value of any of the changing cells.
Allowable decrease	This indicates the change that must be present in the objective coefficient before there is a corresponding decrease in the optimal value of any of the changing cells.

The information provided for each constraint cell is as follows:

Shadow price	This replaces the Lagrange multiplier and indicates the increase of the target per unit increase in the RH side of the constraint equation.
Constraint RH side	This indicates the constraint values specified in the problem.
Allowable increase	This indicates the change that may be present in the constraint RH side value before there is an increase in the optimal value of any of the changing cells.
Allowable decrease	This indicates the change that may be present in the constraint RH side value before there is a decrease in the optimal value of any of the changing cells.

8.6 Nomenclature for Answer Reports

Target cell	This is given as entered in the Solver dialog box.
Status column, binding	This indicates that the final cell value equals the constraint value.
Status column, not binding	This indicates that the final cell value is different from the constraint value but that the constraint condition is met.
Slack column	This indicates the difference between the final cell value at solution and the constraint value. For all binding cells, the slack is zero.

8.7 Nomenclature for Limits Reports

Lower limit	This indicates the smallest value that a changing cell may have while holding all the other changing cells constant, still satisfying all the constraints.
Upper limit	This indicates the largest value that a changing cell may have while holding all the other changing cells constant, still satisfying the constraints.
Target result	This indicates the value of the target cell when the changing cell takes on its lower or upper limit.

Problems

8.1 Maximize or minimize the following y functions using Excel Solver:

a. $y = 3x_1 + 4x_2$

subject to the constraints

$x_1 + x_2 \leq 5$

$x_1 + 2x_2 \leq 10$

$x_1 \geq 0$

$x_2 \geq 0$

b. $y = x_1 + 3x_2$

subject to the constraints

$x_1 + x_2 \leq 3$

$2x_1 + 3x_2 \leq 11$

$x_1 \geq 0$

$x_2 \geq 0$

c. $y = 5x_1 + x_2$

subject to the constraints

$x_1 + x_2 \leq 7$

$3x_1 + 2x_2 \leq 10$

$x_1 \geq 0$

$x_2 \geq 0$

d. $y = 50x_1 + 40x_2$

subject to the constraints

$x_1 + x_2 \leq 900$

$6x_1 + x_2 \leq 9000$

$x_1 \geq 0$

$x_2 \geq 0$

e. $y = 15x_1 + 20x_2$

subject to the constraints

$2x_1 + 5x_2 \leq 700$

$x_1 + 5x_2 \leq 8000$

$x_1 \geq 0$

$x_2 \geq 0$

8.2 Using Excel Solver, maximize or minimize the following y functions subject to the given constraints:

a. $y = x_1 + x_2$

subject to the constraints

$3x_1 + 2x_2 \leq 30$

$x_1^2 + x_2^2 \geq 36$

$x_2 \leq 10$

$x_1 \geq 0$

$x_2 \geq 0$

b. $y = x_1 + x_2$

subject to the constraints

$x_2 \leq 14$

$x_1^2 + x_2^2 \geq 49$

$x_1 \geq 0$

$x_2 \geq 0$

c. $y = x_1 + x_2$

subject to the constraints

$x_2 \leq 11$

$2x_1^2 + 3x_2^2 \geq 100$

$x_1 \geq 0$

$x_2 \geq 0$

d. $y = x_1 + x_2$

subject to the constraints

$x_2 \leq 13$

$2x_1^2 + 3x_2^{1.8} \geq 90$

$x_1 \geq 0$

$x_2 \geq 0$

e. $y = x_1^{1.5} + x_2^{1.5} + 3$

subject to the constraints

$x_2 \leq 15$

$x_1^2 + x_2^2 \geq 36$

$x_1 \geq 0$

$x_2 \geq 0$

8.3 Minimize:

$$y = 10,000x_1^2 + 170x_2$$

subject to the constraints

$$70/x_1^2 \leq K \quad \text{for } K = 1, 2, 3, 4, 5, 6, 7, 8$$

$$x_2 - 160,000/x_1^5 = 0$$

What do you conclude from these calculations?

8.4 Maximize the function:

$$y = \left[\ln(x_2/x_1)/0.05 + 0.2/x_2 \right]^{-1}$$

subject to the constraints

$$x_2 \geq 0$$

$$x_1 - K \geq 0$$

where K is a constant that takes on the values 0.001, 0.005, 0.007, 0.015, and 0.02. What do you conclude from these calculations?

9

Pivot Tables

9.1 Introduction

Pivot tables are most frequently employed to rearrange, regroup, modify, or analyze data appearing in lists or tables. The name *pivot* comes from the fact that the tables are devices for "twisting" or "pivoting" data around a central core of information. Examples of their use include rearranging and presenting sales or manufacturing data by region or product groups and subgroups, profit and supply schedules for different items, or other applications primarily concerned with business situations. Because of the way they operate, pivot tables may also be used effectively to arrange, modify, and present engineering data from different perspectives. They also offer graphing and charting features that can be quite useful.

As with some other topics we have discussed, we will not present a general discussion of pivot tables but will rely on specific examples to illustrate their utility and capabilities. Important terms will be defined and different Windows options will be explained as the examples progress or as the situation warrants. Readers may expand their knowledge of the topic by consulting the Help feature of Excel or searching a topic using their favorite Internet search engine, both of which we may suggest from time to time.

Creating a pivot table in the latest version of Excel is available from the INSERT ribbon bar, similar to creating a Chart. The examples below highlight a number of features of pivot tables and demonstrate how easy they are to create and how useful they can be.

Example 9.1: Analysis of Performance of Air-Conditioning Unit

The performance data of the air-conditioning unit of Example 6.13 are displayed in columns A, B, and C of Figure 9.1. Column D has been added to furnish data for the electric work input W (in units of kW) for the unit. The terms in the table are thus defined as

T_{ew} = inside wet bulb temperature, °F
T_c = outside air temperature for unit, °F
Q_{ew} = cooling performance of the unit, kBtu/h
W = electric power input, kW

In addition, two other parameters are important:

$$EER = Q_{ew}/W = \text{energy efficiency ratio} \tag{9.1}$$

$$Q_h = Q_{ew} + 3.413W = \text{heat dissipated by unit, kBtu/h} \tag{9.2}$$

We will use the pivot table feature of Excel to analyze these parameters. First, we will examine just T_{ew}, T_c, and Q_{ew} (somewhat repeating Example 6.13) to show the pivot table operational features while keeping the presentation very simple.

A pivot table is created using the following procedure:

1. Create a spreadsheet with the data shown in Figure 9.1. Columns A, B, and C will be used for initial creation of the pivot table.
2. Create a pivot table using the INSERT/TABLES/Pivot Table. The Create Pivot Table Wizard window will appear as shown in Figure 9.2. Note that the data range selected is cells A3:C19. For this example, we choose to create the pivot table in a separate worksheet.
3. Click OK. A new worksheet is added to the workbook and the screen now appears as shown in Figure 9.3. At the right side of the window appear the three data fields (Tew, Tc, and Qew) that we wish to analyze or arrange. Later, we will add W, EER, and Qh as data fields for analysis.
4. We choose to display the respective fields by dragging Tew to the row location, Tc to the column location, and Qew to the body of the table (Data Field). These choices are shown in Figure 9.4. Note that the term Sum of Qew appears. By default, Excel displays the sum of all values of Qew for a set of Tew and Tc values. There is only one value of Qew for each set in the data, so this default is the same as listing the values of Qew. For now, ignore the Sum nomenclature. Later, we will see that it is possible to perform other functional manipulations of the data.

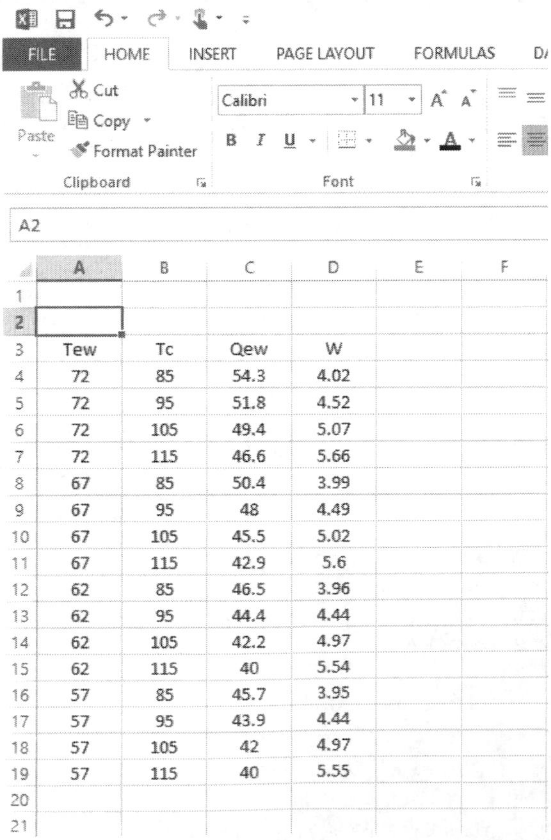

	A	B	C	D	E	F
1						
2						
3	Tew	Tc	Qew	W		
4	72	85	54.3	4.02		
5	72	95	51.8	4.52		
6	72	105	49.4	5.07		
7	72	115	46.6	5.66		
8	67	85	50.4	3.99		
9	67	95	48	4.49		
10	67	105	45.5	5.02		
11	67	115	42.9	5.6		
12	62	85	46.5	3.96		
13	62	95	44.4	4.44		
14	62	105	42.2	4.97		
15	62	115	40	5.54		
16	57	85	45.7	3.95		
17	57	95	43.9	4.44		
18	57	105	42	4.97		
19	57	115	40	5.55		
20						
21						

FIGURE 9.1

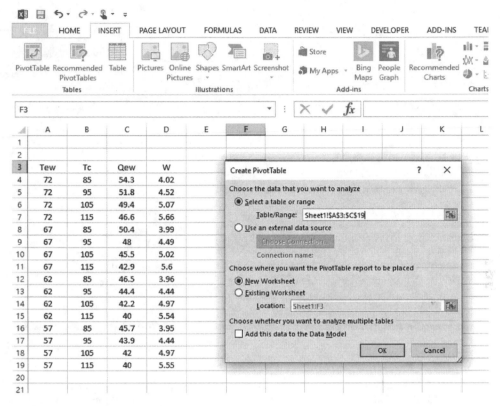

FIGURE 9.2

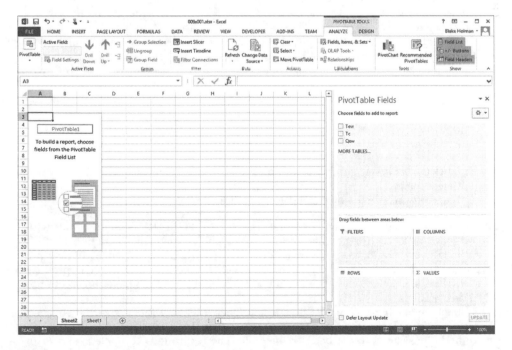

FIGURE 9.3

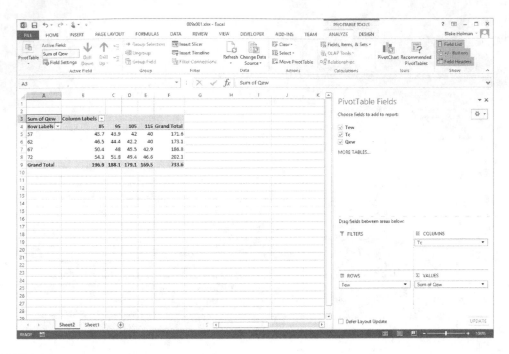

FIGURE 9.4

Note that Tew appears generically as Row Labels and Tc appear generically as Column Labels. To rename these cells more descriptively, double click on the Row Labels and type in a new label. Similarly, perform the same step for Column Labels and type in a new label.

Note also the appearance of the Grand Total field, which sums the row and column values of Qew. These sums are of no interest and can be removed by right clicking on the Grand Total and selecting Remove Grand Total. The revised version of the pivot table appears as in Figure 9.5 with rows and columns labeled and Grand Totals removed.

At this point, we note that the placement of the Tew, Tc, and Qew fields may be altered to change the way the data are displayed.

In addition to a pivot table, the reader may want to display the data in a chart next to the table. Figure 9.6 is obtained by the following:

1. Click the upper-left corner of pivot table (Sum of Qew). This activates the data table.
2. Select INSERT/CHARTS/Line Chart/Stacked Line with Markers. See Chapter 3 for information on format and cosmetic adjustments for Charts.

An alternate display of the pivot table data is shown in Figure 9.7 as an area chart with 3-D visual effects. Again, see Chapter 3 for format and cosmetic instructions.

Although one would normally treat Tew and Tc as the entry variables to obtain a value of Qew, a situation might arise in which values of Tew and Tc that will produce a specific performance cooling value of Qew are sought. The pivot table can accommodate this type of request.

First, the pivot table of Figure 9.4 is activated. Then, the Pivot Table Wizard is engaged to reassign the fields, with Qew assigned to the row field, Tew assigned to the column field, and Tc assigned to the body of the table. The resulting pivot table is shown in Figure 9.8. Note that there are several empty cells in the body of the table.

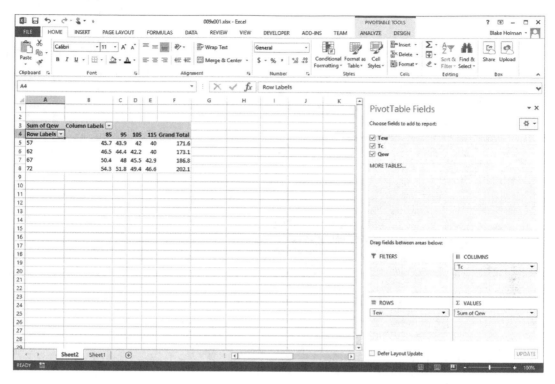

FIGURE 9.5

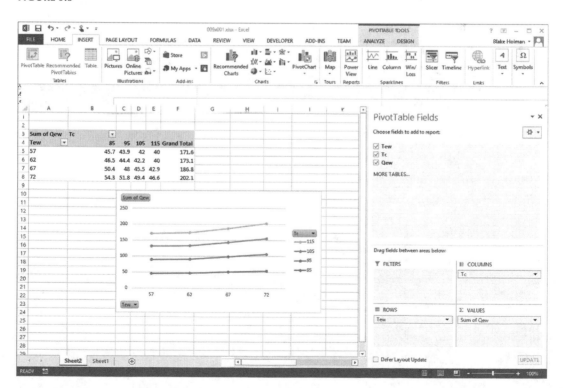

FIGURE 9.6

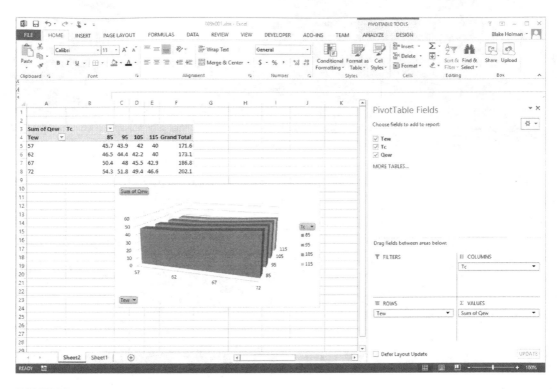

FIGURE 9.7

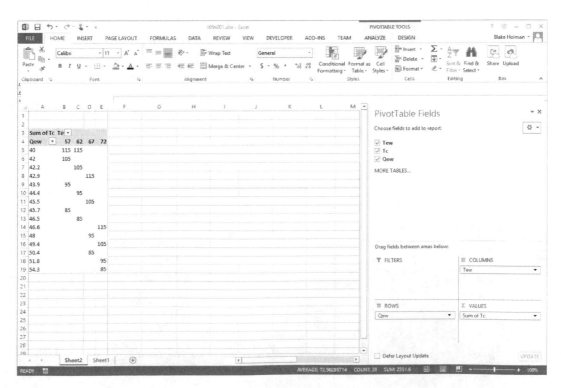

FIGURE 9.8

Example 9.2: Introducing Calculated Data Fields in Pivot Tables

In this example, we see how to enter the additional data fields W, EER, and Q_h into the pivot table of Example 9.1. EER and Qh are "calculated" from the values of Qew and W. The calculations could be performed on the worksheet, but we will see how they may be executed in the pivot table to illustrate that particular feature.

First, all four columns of Figure 9.3 are activated and then the Pivot Table Wizard is engaged to produce a new worksheet for the pivot table. The data elements are dragged to their appropriate locations resulting in the display of Figure 9.9. Note that both Qew and W are displayed in the body of the table as functions of Tew and Tc.

To insert the calculated values of EER and Qh as data fields, you must first create the calculated fields for those values. To do this, perform the following steps:

1. Activate the Pivot Table to display two new menus on the ribbon bar under the heading of PIVOTTABLE TOOLS.
2. Navigate to PIVOTTABLE TOOLS/ANALYZE/CALCULATIONS/Fields, Items & Sets and click on Calculated Field. Enter EER under Name and enter the formula = Qew/W, producing the result in Figure 9.10.
3. Click Add and repeat step 2 for adding Qh with the formula = Qew + 3.413 × W.
4. All four data fields are in the Pivot Table Wizard to produce the result shown in Figure 9.11. The fields may be moved in and out of the pivot table as desired.

Example 9.3: Addition of Regression Results to Pivot Table

It is a simple matter to add the results of the linear regression analysis of Example 6.13 to the pivot tables for Qew presented in the preceding text. The results of the

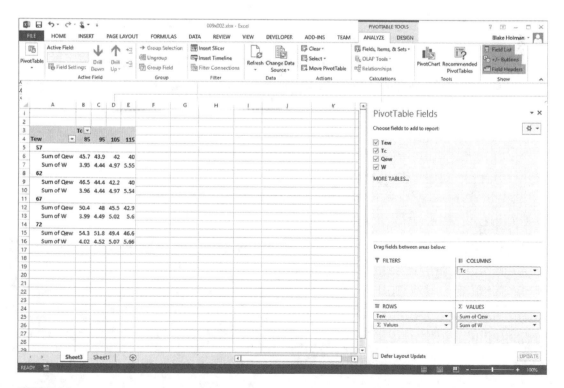

FIGURE 9.9

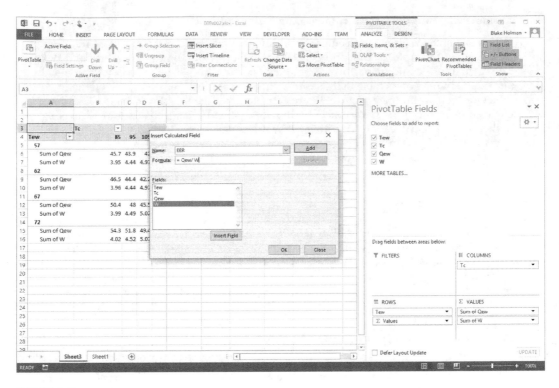

FIGURE 9.10

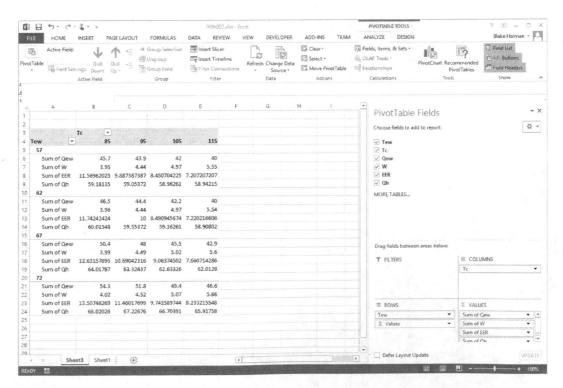

FIGURE 9.11

regression for Qew as a function of Tew and Tc and written in format for entry in the pivot table become:

$$Qewr = 0.526 \times Tew - 0.228 \times Tc + 34.723 \qquad (9.3)$$

where we now use the symbol Qewr to designate the result of the linear regression analysis. The deviation from the original tabular values may be written as DevQewr and

$$DevQewr = Qew - Qewr \qquad (9.4)$$

or, the percent deviation as

$$\%DevQewr = (DevQewr/Qew) \times 100 \qquad (9.5)$$

The formulas in the preceding equations may be entered as calculated fields in the pivot table of Example 9.1. The resulting pivot table for data fields of Qew, Qewr, DevQewr, and %DevQewr is displayed in Figure 9.12. Graphical presentations of the results are easily obtained as outlined in the previous examples.

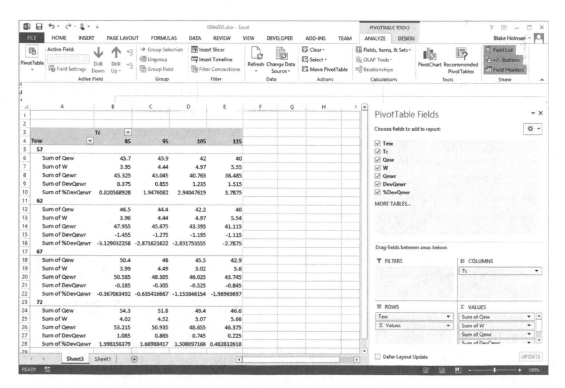

FIGURE 9.12

9.2 Other Summary Functions for Data Fields

As noted in Example 9.1, the Excel default pivot table displays sums of the data items for each set of row and column values. When there is only one value for each set, as in the previous examples, the sum action is of no consequence. If multiple values are present in the data, Excel provides other choices for the data presentation in addition to the sum function. These choices may be displayed and set by the following procedure:

1. Begin with the Pivot Table created as shown in Figure 9.4.
2. Right click the Sum of Qew field in the Pivot Table, and select Value Field Settings from the menu. The result appears as shown in Figure 9.13. The function choices for summary of the data are as follows:

   Sum

   Count

   Average

   Max

   Min

   Product

   Count numbers

   StdDev

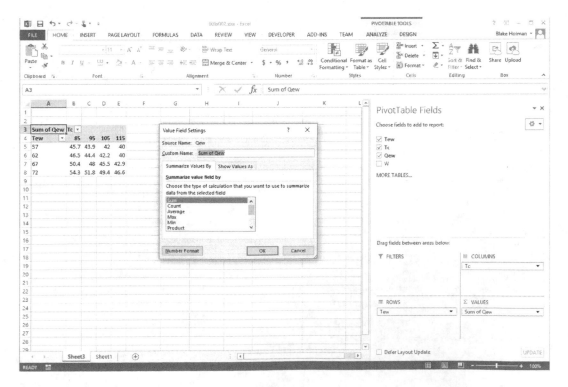

FIGURE 9.13

StdDevp

Var

Varp

3. Row and column fields may also be manipulated in different ways by right click-ing on one of the row or column values to select the Field Settings menu item and then selecting from a similar set of choices. The result of such action for the Tew rows is shown in Figure 9.14. In the case of rows and columns, the Subtotals are obtained in contrast to the Grand Totals described previously. These Subtotals may be deleted from the pivot table presentation by clicking None as a choice in Figure 9.14. Selection of Automatic displays the Subtotals.

Example 9.4: Pivot Table for Basic Financial Functions

As another example, we now show how pivot tables may be employed to calculate and present the basic financial functions of Section 7.3. The formulas are repeated below for convenience:

$$fv = (1+I)^n \tag{9.6}$$

$$pmt = I/\left[1-(1+I)^{-n}\right] \tag{9.7}$$

$$pv = \left[1-(1+I)^{-n}\right]/I \tag{9.8}$$

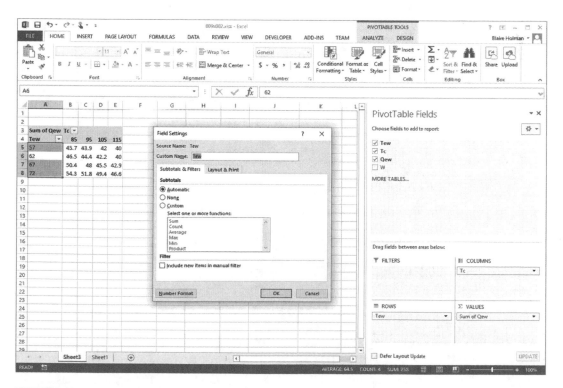

FIGURE 9.14

$$\mathrm{fvp} = \left[(1+\mathrm{I})^{\mathrm{n}} - 1\right]/\mathrm{I} \tag{9.9}$$

where we now use the symbol fvp to designate the future value of uniform periodic payments. The formulas may alternately be expressed in terms of fv calculated from Equation 9.6 as

$$\mathrm{pmt} = \mathrm{I}/(1 - 1/\mathrm{fv})$$

$$\mathrm{pv} = (1 - 1/\mathrm{fv})/\mathrm{I}$$

$$\mathrm{fvp} = (\mathrm{fv} - 1)/\mathrm{I}$$

A table is set up, as shown in Figure 9.15, for I values of 3%, 4%, 5%, and 6% and n values of 5, 10, 15, 20, 25, and 30 periods in columns A and B. The values of fv are calculated in column C using Equation 9.6. The table is then activated and the Pivot Table Wizard is engaged to produce the initial pivot table as shown in Figure 9.16. Subsequently, the other functions are added as calculated data fields, using the procedure outlined in Section 9.1. The pivot table displaying all the resulting data is shown in Figure 9.17. Pivot

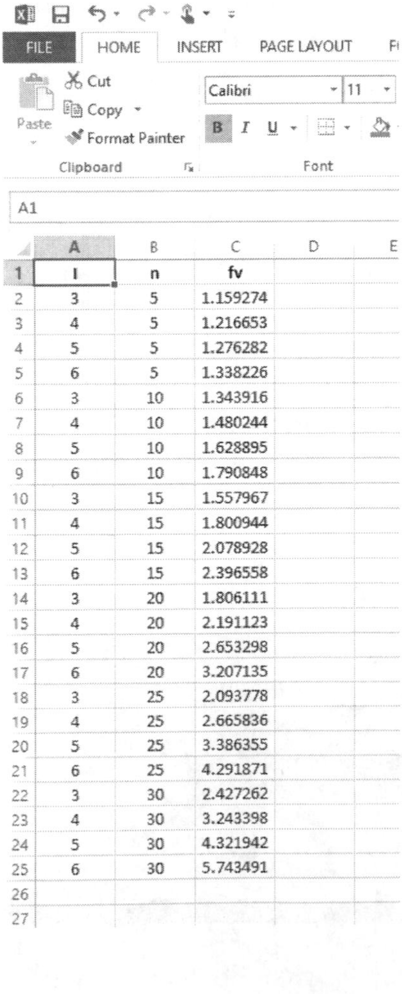

	A	B	C
1	I	n	fv
2	3	5	1.159274
3	4	5	1.216653
4	5	5	1.276282
5	6	5	1.338226
6	3	10	1.343916
7	4	10	1.480244
8	5	10	1.628895
9	6	10	1.790848
10	3	15	1.557967
11	4	15	1.800944
12	5	15	2.078928
13	6	15	2.396558
14	3	20	1.806111
15	4	20	2.191123
16	5	20	2.653298
17	6	20	3.207135
18	3	25	2.093778
19	4	25	2.665836
20	5	25	3.386355
21	6	25	4.291871
22	3	30	2.427262
23	4	30	3.243398
24	5	30	4.321942
25	6	30	5.743491

FIGURE 9.15

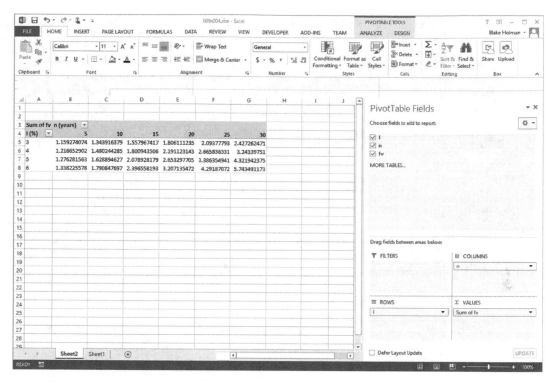

FIGURE 9.16

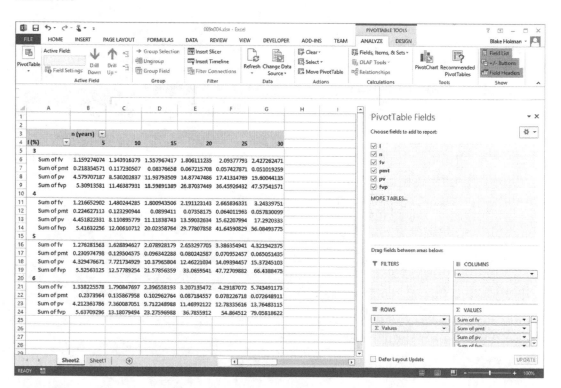

FIGURE 9.17

tables for the fv and fvp functions are shown in Figures 9.18 and 9.19, respectively, with line graphs (no data markers) displayed. In both cases, the graphs are presented alternately with n and I as the abscissa.

Although the financial functions could have been calculated directly on a worksheet and subsequently displayed in the respective graphs, the use of pivot tables offers increased flexibility of presentation possibilities, with much less fuss about switching columns and rows, clicking and dragging to select data for charts, etc.

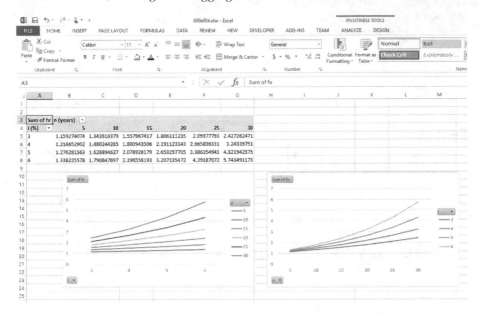

FIGURE 9.18

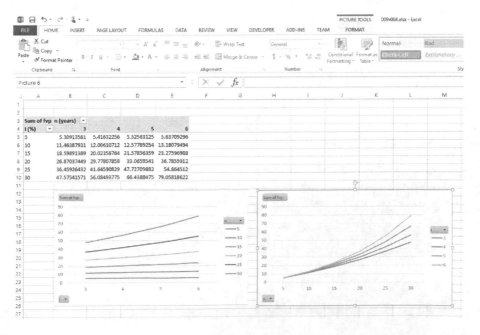

FIGURE 9.19

9.3 Restrictions on Pivot Table Formulas

Several restrictions are placed on the way formulas are constructed to enter calculated data fields in pivot tables. For engineering work, perhaps the most restrictive requirement is the fact that worksheet functions (EXP, SQRT, etc.) may not display a cell reference or defined names as parameters in the syntax statements for the functions. Array references as used in the function BESSELJ(n, x) are also not permitted. In many cases, the restrictions can be circumvented by setting up the formulas in terms of the data fields of the pivot table as was done in Examples 9.2 through 9.4. The reader should consult Excel Help under "Syntax for calculated field and item formulas in pivot tables" for additional information.

9.4 Calculating and Charting Single or Multiple Functions $\partial(x)$ vs. x Using Pivot Tables

In accordance with our previous discussion, a procedure for calculating functions $\partial(x)$ may be described as follows:

1. Open an Excel workbook. Column A will be used for values of x. The initial value of x is assigned in cell C1. The increment in x, Dx, is assigned in cell E1. Enter the formula = C1 in cell A2. Enter the formula = A2+$E$1 in cell A3. Drag-copy cell A3 for as many rows as desired for the calculations: Several hundred rows are suggested to accommodate small increments in x. The upper-left portion of sheet 1 will appear as shown in Figure 9.20 for x1 = 0 and Dx = 0.2.
2. Assign the desired values of x1 and Dx for the calculations.
3. Insert a new pivot table at cell G1 (in the existing worksheet), where the data for x is in both the rows and the body of the pivot table. Remove the grand totals.
4. The simple data for f(x) = x in the pivot table are activated and plotted in a line graph at C16:J30. The resulting worksheet is shown in Figure 9.21.

Additional functions of x are now entered in the pivot table by the following procedure:

1. Activate the pivot table by clicking the upper-left corner.
2. Add calculated fields to the data set by navigating to PIVOTTABLE TOOLS/ ANALYZE/CALCULATIONS/Fields, Items & Sets and clicking on Calculated Field. Name new functions under Name: and insert the associated formula for the name in the line provided. Several functions are shown in Figure 9.22 as typical additions. For example, the formula for sin2x would be entered as = SIN(2 * x). Click Add. Insert as many functions as desired, and then click OK. As each new field is added, the pivot table will be updated (Figure 9.23).
3. Functions may be graphed singly or in multiples by activating the pivot table and using Chart Wizard. Multiple function charts may become rather cluttered, and discretion is advised. A line graph for sin(nx) for n = 1–5 is shown in Figure 9.24. Note that the curves are not at all smooth as one would expect of a plot of a sine curve. This raises a significant limitation of pivot tables in Excel 2016 as described in Section 9.4.1.

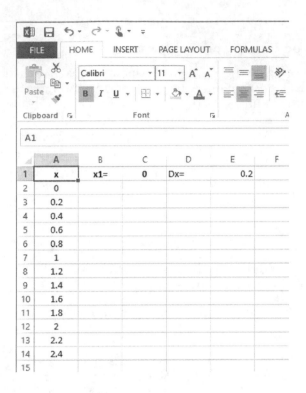

FIGURE 9.20

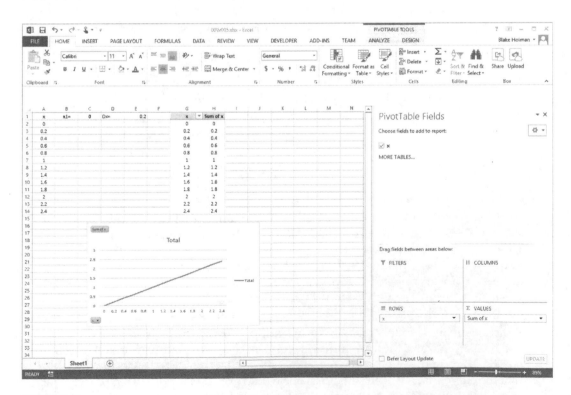

FIGURE 9.21

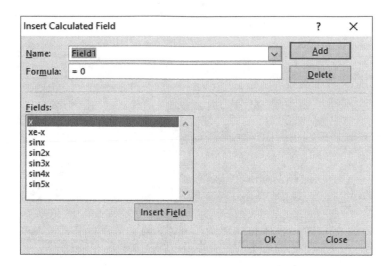

FIGURE 9.22

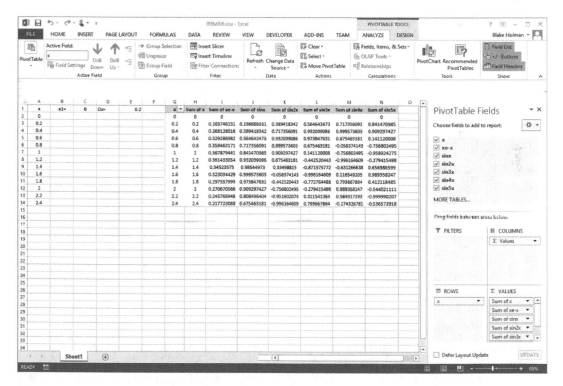

FIGURE 9.23

9.4.1 Working around Charting Limitations of Excel 2016—Scatter Charts

Unfortunately, Excel 2016 has a significant limitation in that it will not allow the user to create a scatter chart of data created using a pivot table. Attempting to do so results in the error message displayed in Figure 9.25. This is a limitation for reasons related to data mismatch between what a pivot table provides and what a scatter chart requires.

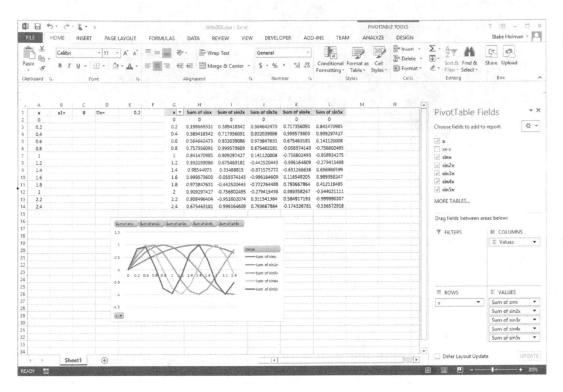

FIGURE 9.24

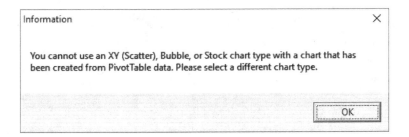

FIGURE 9.25

Like many things in life, though, there is an alternative approach to creating a scatter chart from pivot table data. Consider the line graph of Figure 9.24. A simple method for creating a scatter chart from data in a pivot table is to copy the data to another area of the worksheet or to a different worksheet and then create the scatter chart from the copied data. To accomplish this with the data in Figure 9.24, perform the following:

1. Highlight the data in the pivot table by selecting cells G1:L14 in Figure 9.24.
2. Copy the selected data using the CTRL+C key sequence.
3. Select cell G17 and right click to paste the values copied from the pivot table.
4. Select the copied cells (G17:L30) and then insert a type 3 scatter chart.

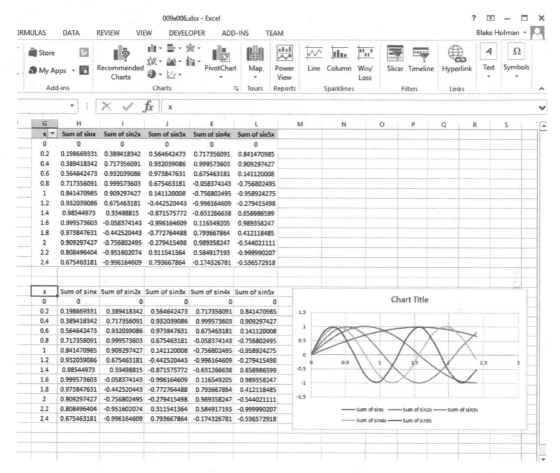

FIGURE 9.26

The resulting worksheet is shown in Figure 9.26.

There are other methods for working around this limitation in Excel 2016, one of which is the use of Power Query, which is discussed in Section 10.6 of this text.

9.5 Calculating and Plotting Functions of Two Variables

We have already seen in Example 9.4 the calculations for functions of two variables; in this case, financial functions are functions of I and n. Because pivot tables cannot accept array formulas, it may sometimes be preferable to use the DATA/FORECAST/What-If-Analysis/ Table command for such calculations, as described in Section 2.15. However, pivot tables offer the advantage of easy entry of several functions into the table and quick subsequent manipulation of these functions. Taking the two entry variables as x and y, we wish to calculate the $\partial(x, y)$ values. Again, because of the non-array entry requirements, all combinations of the variables x and y must be present in the entry fields. Note in Figure 9.15 how this requirement was met for the variables I and n. We consider another example—one from mechanical vibrations. The functions to be calculated are as follows:

$$a = \text{amplitude function} = x^2 / \left[(1 - x^2)^2 + (2xy)^2 \right]^{0.5} \tag{9.10}$$

$$a2 = \text{second amplitude function} = a/x^2 \tag{9.11}$$

$$p = \text{phase shift angle} = \tan^{-1} \left[2xy / \left(1 - x^2 \right) \right] \tag{9.12}$$

where

$x = \omega/\omega_n = $ frequency ratio
$y = c/c_c = $ damping ratio

A worksheet is opened as shown in Figure 9.27 and the values for x and y are entered in columns A and B for x = 0–2 in 0.1 increments and y-values of 0.25, 0.5, 0.7, and 1.0. The formulas corresponding to the three equations for a, a2, and p are entered in the pivot table and the calculated results are plotted as shown in Figure 9.28. Note that the type 3 scatter charts in Figure 9.28 were created using the work-around discussed in Section 9.4.1.

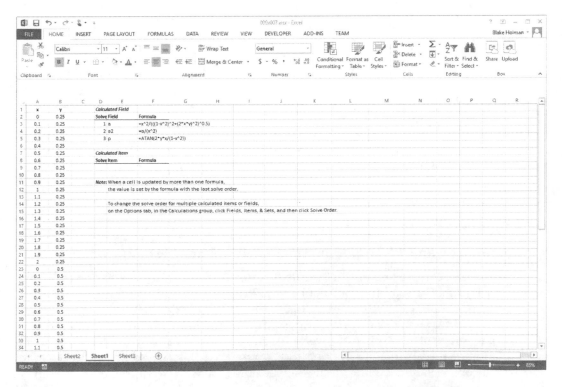

FIGURE 9.27

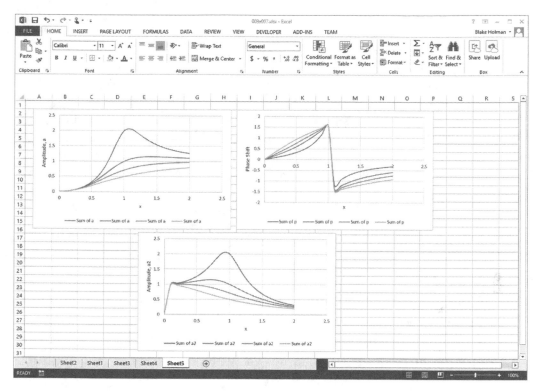

FIGURE 9.28

Problems

9.1   The following function occurs in an engineering design application:

$$e = 1 - \exp\left\{\left[\exp(-NnC) - 1\right]/nC\right\}$$

where

$$n = N^{-0.22}$$

Set up a pivot table to calculate values of e as a function of N and C for the ranges:

$$0 < N < 5 \text{ and } 0 < C < 1$$

Then, provide a graph for several values of these variables. Plot as $e = \partial(N)$ for different values of C and also as $e = \partial(C)$ for selected values of N.

9.2   Set up a pivot table similar to that in Example 9.3 but for the exponential regression of Example 6.14.

9.3   Set up a pivot table similar to that of Example 9.3 but based on the combined regression analysis of Example 6.15.

9.4   Set up a pivot table to calculate P/B from Equation 7.17b for I = 3%, 5%, 7%, 9%, and 12% with values of n = 5, 10, 15, 20, and 25 periods. Select values of C as appropriate. Note that values of C = I will result in division by zero, but selecting I = 3 and C = 2.99 will not produce an error. Provide a graph of P/B for one value of n.

9.5   Perform a second-order polynomial fit for the W(kW) data of Example 9.1 using a single value of T_{ew} = 57°F along with the four given values of T_c. Obtain the following:

$$W = aT_c^2 + bT_c + c \text{ for } T_{ew} = 57°F = \text{constant}$$

On a hot summer day in Austin, Texas, the temperature T_c varies according to

$$T_c = 93.5 + 8.5 \sin(0.2618t - 2.8798)$$

where t is the time measured in hours from midnight. Using the combination regression obtained for Qew in Example 6.15, construct a pivot table that will

a. Present values of Qew(kBtu/h) as a function of t

b. Present values of W(kW) as a function of t

c. Sum the total cooling Qew(kBtu) and energy input W(kWh) over a 24-h period

d. Display values of T_c as a function of t

Using this information, plot Qew(kBtu/h), W(kW), and T_c as functions of t in hours.

9.6   The cooling load that the air-conditioning unit in Problem 9.5 must accommodate variation with T_c according to the relation:

$$Qload = \text{Const} \times [0.03704(T_c - 75) + 0.2332]$$

Assume that the unit is oversized such that it can deliver 10% more cooling than the maximum needed under the most severe temperature conditions, i.e., for T_c = 102°F at 5 pm. Devise additions to the pivot table of Problem 9.5 that will calculate and display a duty cycle factor defined by

$$F = Qload(t)/Qew(t)$$

Plot the values of F as a function of t.

9.7   As the air-conditioning unit in Problems 9.5 and 9.6 is oversized, a proposal is made to store the excess capacity by suitable means, namely, by storing chilled water or making it available for other applications. Provide additions to the pivot table that will display the hourly Btu of excess capacity that will be available.

9.8   The particle displacement amplitude for a "standing" sound wave may be described by the relation:

$$a = \sin(2\pi x/\lambda)\times(\cos 2\pi ct/\lambda + const)$$

where

    x = spatial location, m
    λ = wavelength of the sound, m
    c = acoustic velocity, m/sec
    t = time, sec

Devise a pivot table to display and graph a suitable range of values of the amplitude function for $\lambda = 0.2$ m and $c = 344$ m/s.

9.9   The steady-state amplitude response of a first-order system to an impressed frequency ω is given by

$$a(t) = \sin(\omega t - f)/[1+(\omega t)^2]^{1/2}$$

where ϕ is the phase shift angle defined by

$$\phi = -\tan^{-1}(\omega t)$$

Devise a pivot table to display this response. Plot the results on a suitable graph.

9.10 The time-delay behavior of a low-frequency ($\approx$1 cycle per 24 h) thermal wave in a large solid is described by the amplitude response equation:

$$a(x, t) = \exp\left[-(x^{2\pi\omega}/\alpha)^{0.5}\right]\times\sin\left[2\pi\omega t - (x^{2\pi\omega}/\alpha)^{0.5}\right]$$

where

    x is the depth in the solid material, m
    ω is the frequency of the thermal wave at $x = 0$, in cycles/s
    t is the time, s
    α = constant = 7×10^{-7} m^2/s

Construct a pivot table and appropriate charts to display the behavior of $a(x, t)$ over at least two cycles of the impressed wave at $x = 0$. Also, examine the behavior of a time-delay function:

$$\Delta t = (x^2/4\alpha\pi\omega)^{1/2}$$

9.11 A transient cooling problem involves the equation:

$$\theta = 1 - erf(X) - \left[\exp(hx/k + h^{2\alpha}t/k^2)\right]\times\left\{1 - erf\left[X + (h^{2\alpha}t/k^2)^{0.5}\right]\right\}$$
$$= \text{function of }\left[X, (h^{2\alpha}t/k^2)^{0.5}\right]$$

where

$X = (x^2/4\alpha t)^{0.5}$

$\alpha, h, k = $ constants

$x = $ space coordinate

$t = $ time coordinate

Devise a pivot table to display values of θ in the range from 0.01 to 1.0 for ranges of the variables:

$$0 < X < 1.5$$

$$0.05 < (h^{2\alpha}t/k^2)^{0.5} < \infty$$

10

Data Management Resources in Excel

10.1 Introduction

As we have seen in several examples throughout this volume, Excel can hold tables of data for calculations, analysis, and charting purposes. Excel can also hold data for reference and processing purposes. When there is a need to keep data, calculations, analyses, and charts together, worksheets and tables within an Excel workbook make an excellent choice. There are, however, times when the data may not be available for inclusion in an Excel workbook or may be better referenced from an external data source.

In addition to supporting various methods for storing and accessing data, Excel has numerous capabilities for managing and processing data, over and above the analyses and charting capabilities discussed thus far. This chapter aims to outline many of these capabilities and methods, which the reader is then encouraged to employ in his/her day-to-day work.

10.2 Organizing Data in Excel Worksheets and Tables

The simplest way to organize data in Excel is in worksheet rows and columns. Consider a contact list where names and addresses are stored in a worksheet for easy reference. This may look like the data in Figure 10.1.

In this format, the data is very workable—easy to view, and can easily be referenced by additional columns, or by cells in other worksheets whether they are in the same workbook or in other Excel workbooks.

Excel also has the ability to create a table reference by name, rather than relying on worksheet name and row/column information. To create a table named Contacts for the same data, perform the following steps:

1. Highlight the column headings and all the rows of the data which will become the table named Contacts.
2. Press the key sequence CTRL+t or click on HOME/STYLES/Format As Table and pick a style.
3. Upon performing step 2, a dialog box will appear as in Figure 10.2. Verify the information and click OK.

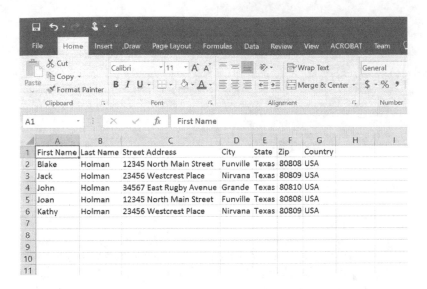

FIGURE 10.1

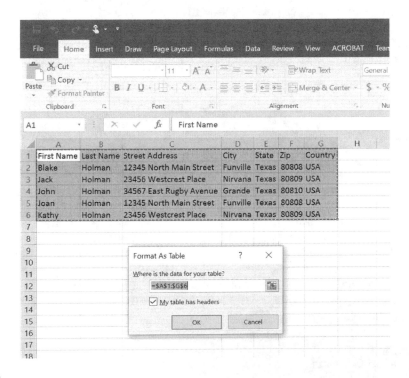

FIGURE 10.2

4. The resulting table will have either the default coloring (by pressing CTRL+t) or the coloring chosen (clicking on HOME/STYLES/Format As Table), and it will have auto-filtering enabled so that the user can sort and filter the columns of the table (Figure 10.3).

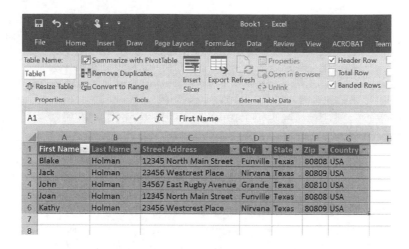

FIGURE 10.3

FIGURE 10.4

5. To give the table the custom name Contacts, place the cursor on a cell in the table and the Table Tools menu will appear on the ribbon bar. Navigate to the Table Tools Design menu and note the upper-left corner of the ribbon bar. It should appear as in Figure 10.3. To change the name of the table to Contacts, type over the term Table 1 with the term Contacts. Navigate away from the table and back to it to ensure that the name is preserved (Figure 10.4).

10.3 Filtering, Sorting, and Using Subtotals

10.3.1 Filtering

Filtering of data is an extremely useful feature, particularly when working with large numbers of rows of data. In the situation of a contact list, imagine wanting to filter names by zip code or country. A brute force method for performing such could be to sort the data by either zip code or country and then hide the rows that don't fit with what you seek. Alternatively, using Excel's built-in filtering capabilities can automatically accomplish these steps.

The simplest way to engage filtering is to use the Auto-Filter facility of Excel. Highlight the column headings and data you want to filter and navigate to DATA/SORT & FILTER and click on Filter. Excel will enable a drop-down filter on each column from which you can perform several actions. These potential actions include the following:

- Sort A to Z (ascending order)
- Sort Z to A (descending order)
- Sort by Color (if colors are enabled and used)
- Clear Filter from {column name}
- Filter by Color
- A Search box for values in that column
- A box that allows the user to select which values to show or hide

Consider the data in Figure 10.1. Enabling filtering on that data set and then selecting the drop-down menu on the Street Address field would appear as in Figure 10.5.

By unchecking the Select All value and then checking the 12345 North Main Street value, the table of data will be reduced to only two rows with 12345 North Main Street as a value for the Street Address column.

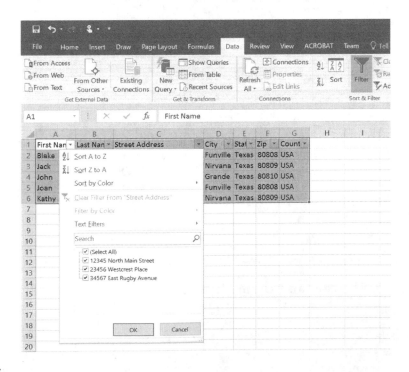

FIGURE 10.5

10.3.2 Sorting

As seen from the menu choices above, sorting can be accomplished using the filtering drop-down menus on each column. If, however, nested sorting (sorting by value A, then value B within value A, etc.) is needed, an alternative approach is needed.

Consider the data in Figure 10.6. Imagine that these data need to be sorted first by Last Name, then by First Name, and then by Middle Initial. To create a nested sort for these data, perform the following steps:

1. Highlight the data to be sorted.
2. Click DATA/SORT & FILTER/Sort to bring up the dialog box in Figure 10.7.
3. Click on the My data has headers checkbox in the upper-right corner.
4. Click the Sort by drop-down box and pick Last Name as the first sort item. Keep the Sort on Values choice and keep the Order of A to Z.
5. Click on Add Level to add a next level sort. An additional row will appear in the dialog box. For that new row, select First Name from the Sort by drop-down box leaving the Sort on and Order options at their default values.
6. Click on Add Level again, this time picking Middle Initial from the Sort by drop-down menu, again leaving Sort on and Order as their default values, resulting in the dialog box in Figure 10.8.
7. Click on OK and the data will rearrange as shown in Figure 10.9.

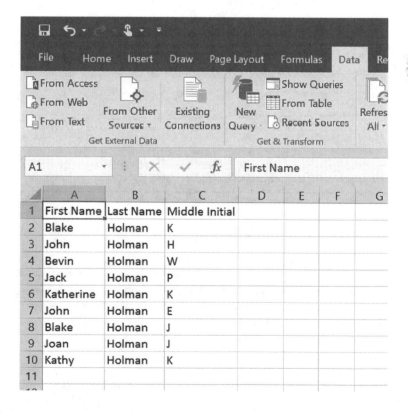

FIGURE 10.6

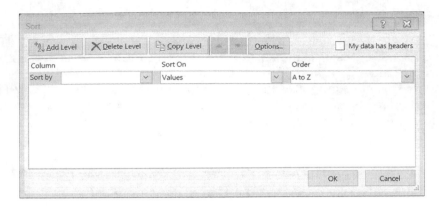

FIGURE 10.7

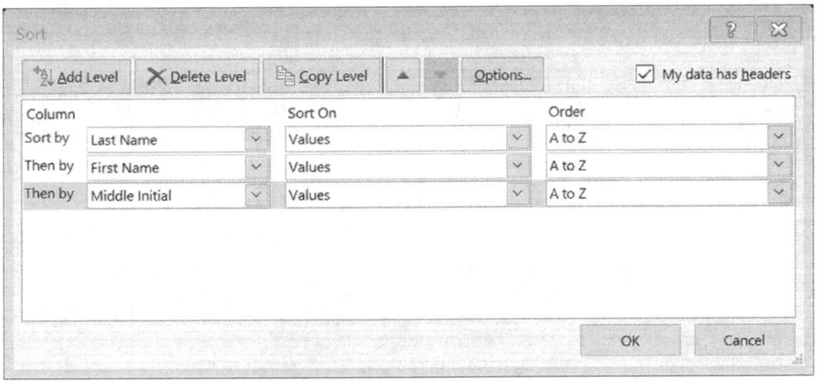

FIGURE 10.8

10.3.3 Subtotals

Subtotals can be used in Excel to perform several operations on subgroups of data. The data must be sorted by a value on which the data are grouped, and then subtotals can be employed on various other data elements. The operations that can be performed on a subgroup of data are Sum, Count, Average, Minimum, Maximum, and Product. Product is a multiplication operation, minimum is the identification of the minimum value of a subgroup, and maximum is the identification of a maximum value in a subgroup.

Consider the data in Figure 10.10. The data consist of a list of spare parts that might be on hand with a present-day IT group for quick repair of problematic devices. Let's say, for example, that management would like to know how many devices are held per category as well as the dollar value of that inventory. Using subtotals for summing the number of items on hand as well as the value of those items would be very handy.

To perform such an operation on this sample data, perform the following steps:

1. Start by highlighting the data and column headings for which subtotals will be created.

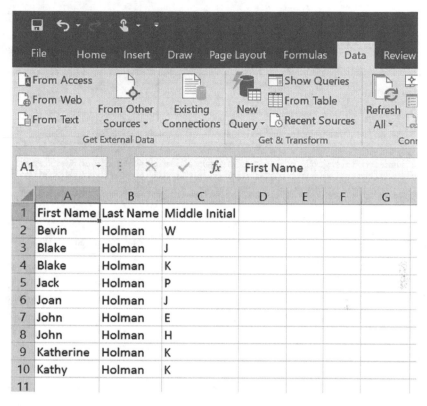

FIGURE 10.9

2. Sort the data by category to ensure that the line items are grouped together by that column. This is accomplished using the DATA/SORT & FILTER/Sort menu item. Make sure that My data has headers is selected and pick the Category column name for the Sort by field.

3. Engage the subtotaling facility via the DATA/OUTLINE/Subtotal menu choice, and the dialog box in Figure 10.11 will appear.

4. Change the At each change in: drop-down value to be Category, and check the box for Number on Hand in the Add Subtotal to: window. Both Number on Hand and Ext. Cost should be checked for subtotals.

5. Click OK to see the result in Figure 10.12. Notice that Excel has inserted subtotal lines for each category and has provided a subtotal of items on hand in that category as well as the values of those items. In addition, a Grand Total is provided at the bottom of the sheet.

6. Note that there are three columns numbered 1, 2, and 3 to the left of the row numbers of the worksheet. These designate the levels of information in the data set. Clicking on the number three shows all details of the data set. Clicking on the number two will show just the subtotals and the Grand Total, and clicking on the number one will show only the Grand Total. Figure 10.13 shows the result of clicking on the number two.

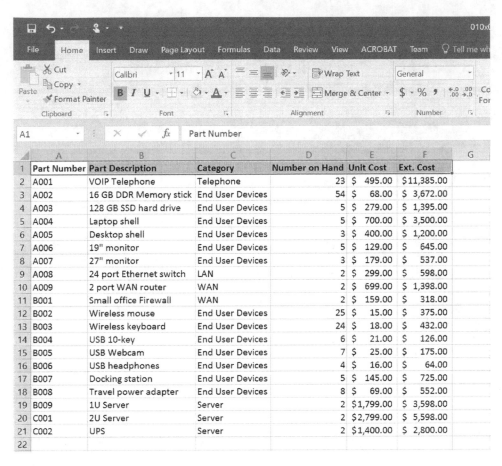

FIGURE 10.10

FIGURE 10.11

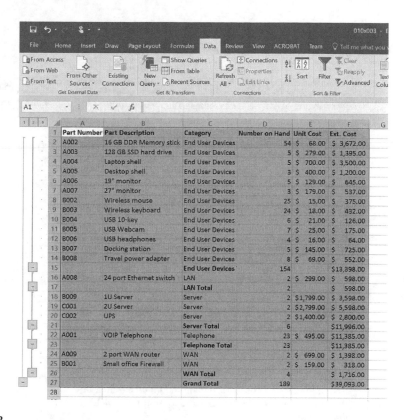

FIGURE 10.12

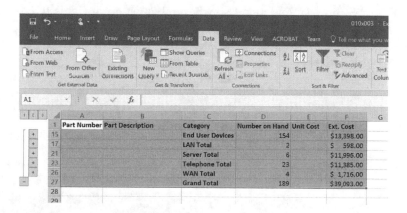

FIGURE 10.13

10.4 Useful Data Functions in Excel

Through many years of using Excel for data processing and manipulation, numerous Text and Lookup/Reference Excel functions show their use and versatility on nearly a daily basis. These very useful functions are shown in Table 10.1. Refer to the Excel Help function (press F1 from within Excel) for more information on each function.

TABLE 10.1

Useful Data Processing and Manipulation Functions

Function	Description
CONCATENATE(), CONCAT()	Used to create one string of characters from several individual strings. CONCAT() is only available in Office 365. Resultant string is limited to 32,767 characters
FIND()	Used to find a string within a string and return the position at which that substring begins. This function is case-sensitive
LEFT()	Used to grab and return the leftmost characters of a string value. The number of characters is specified as an argument to the function
LEN()	Used to return the length of a string value
LOWER()	Used to convert a value to all lowercase characters
MID()	Used to return the middle of a string value given a starting and ending character position
REPLACE()	Used to replace one string with another within a target string
RIGHT()	Used to grab and return the rightmost characters of a string value. The number of characters is specified as an argument to the function
SEARCH()	Used to find a string within a string and return the position at which that substring begins. Unlike the FIND() function, this function is NOT case-sensitive
TEXT()	Used to convert a numeric value to a text value
TRIM()	Used to trim extraneous spaces from a string value except for single spaces between words
UPPER()	Used to convert a value to all uppercase characters
VALUE()	Used to convert a string of numbers from a text value to a numeric value
VLOOKUP()	Used to look up a value in a row of data given a matching value in the first element of the row. Similar to HLOOKUP() except that HLOOKUP() operates on columns rather than rows

Example 10.1: String Manipulation Using Several Functions in Table 10.1

Consider the data set in Figure 10.14. Note that there are several holes in the data set, which the functions in Table 10.1 can easily help us fill. For example,

- The ID field is intended to be the first three letters of the last name with the index appended to it.
- The Full Name is intended to be the piecing together of the First Name, Middle Initial, and Last Name with appropriate spacing.
- The Simple Zip is intended to be the first five characters of the Zip+4 field.
- The Plus 4 is intended to be the last four characters of the Zip+4 field.
- The Area Code is intended to be the first three characters of the Telephone field.

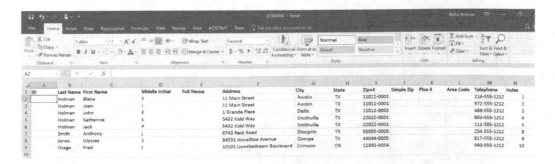

FIGURE 10.14

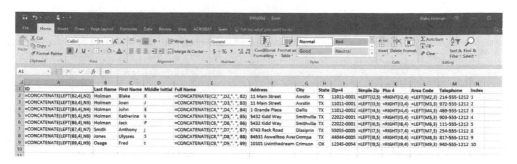

FIGURE 10.15

FIGURE 10.16

The worksheet in Figure 10.14 is put into formula display mode, which you will recall from previous chapters is accomplished via the CTRL+` key sequence. Using several of the functions from Table 10.1, Figure 10.15 shows how these functions can be used in the open columns of data to accomplish the intentions noted above. Figure 10.16 shows the final results by switching back to displaying values from formulas using the CTRL+` toggle key sequence.

The reader is left to add a column for cleaning up the middle initials, making them uppercase, and copying and pasting them back into the proper place. Observe how the Full Name changes when this cleanup is complete.

10.5 Connecting Excel to External Data

When a data source is available for analysis outside of Excel, it is convenient to be able to reference that data source without having to type the data into a worksheet or rebuild the data otherwise. An example of an external data source might be an MS Access database to which a data logging tool has recorded experimental data. Similarly, a Central Plant management system may store numerous data readings on temperatures and water flow throughout the system, which warrant analysis. Excel provides quite a bit of flexibility in connecting to such data sources and accessing the data for analysis, charting, or processing purposes.

10.5.1 General Concepts

Excel has the capability of connecting to data outside of the current workbook and maintaining a reference that such data are external and may need to be refreshed periodically. The external data can be in many forms: an Excel workbook, an MS Access database, an MS SQL Server database, a MySQL database, or one of many others.

The idea is to configure an Excel worksheet to look outside itself for data and show the data in the current worksheet for analysis or manipulation purposes. In the discussions that follow, it is assumed that the reader has a basic understanding of databases and tables in the context of typical database systems. For an in-depth understanding of databases, the reader is encouraged to explore any number of other sources including Microsoft.com and mysql.com.

10.5.2 Connecting to MS Access

Consider a building security system that is logging badge swipes of employee badges as employees enter and exit the doors of their employer's building. Let's say that the logging system has extremely poor reporting capabilities and the company's executive management would like a report of the last 100 people to enter the facility. The logging system stores its data in an MS Access database.

Say that the name of the MS Access database is SecuritySystem.accdb and the table in the database that is home to the badge swipes is named BadgeSwipes. To configure MS Excel to connect to this sample database, perform the following steps:

1. Open a new Excel workbook to Sheet1.
2. Click on DATA/GET EXTERNAL DATA/From MS Access.
3. Locate the file system path to the MS Access database file SecuritySystem.accdb, highlight that filename, and click Open.

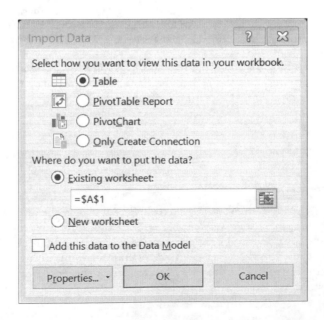

FIGURE 10.17

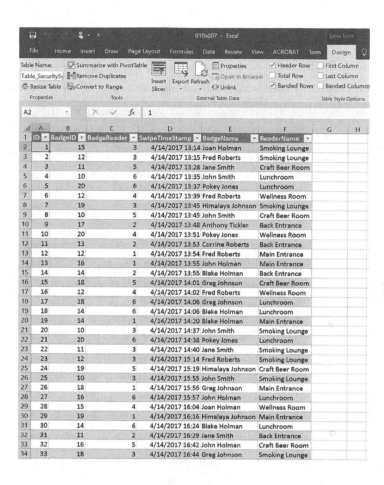

FIGURE 10.18

4. The dialog box in Figure 10.17 will appear. The four options enable the user to

   a. Create a table within the workbook that has a copy of the data from the Access database (and place it in the current or a new worksheet)

   b. Proceed directly to create a PivotTable associated with the data (and place it in the current or a new worksheet)

   c. Proceed directly to create a PivotChart associated with the data (and place it in the current or a new worksheet)

   d. Simply connect to the data without any other operation

5. In our case, accept the settings as they are in Figure 10.17 and click OK. The result appears as in Figure 10.18.

10.5.3 Connecting to MS SQL Server

Consider the situation in Figure 3.6b, except the data are stored in the cloud on an MS SQL Server running at Microsoft (Azure). Assume that the reader has a login and password capable of logging into and reading data from this SQL Server and that the SQL Server firewall has been set up properly to accept connections from the reader's computer. Chapter 11 covers more information about cloud-based services.

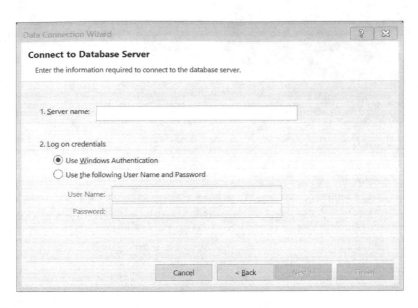

FIGURE 10.19

Say that the hostname of the SQL Server is secsys.database.windows.net. Additionally, the database on the server that holds the data is called ExpData and the table in the database that is home to the data is named Experiment1. To configure MS Excel to connect to this table, perform the following steps:

1. Open a new Excel workbook to Sheet1.
2. Click on DATA/GET EXTERNAL DATA/From Other Sources/From SQL Server.
3. A dialog box like that shown in Figure 10.19 will appear.
4. Enter the SQL Server name, indicate the login credentials being used and click Next.
5. The Data Connection Wizard dialog box will appear. Under Select the database that contains the data you want, select ExpData from the drop-down choices. Doing so will display the Experiment1 table in the list as shown in Figure 10.20.
6. Highlight the Experiment1 table name and click Next; then, click Finish on the Save Data Connection File and Finish screen.
7. A dialog box similar to that in Figure 10.17 will appear. Click OK and the result will look similar to Figure 10.21.

10.5.4 Connecting to MySQL

Consider the data in Problem 3.1, except the data are stored in the cloud on a MySQL Server running at Amazon Web Services. Assume that the reader has a login and password capable of logging into and reading data from this MySQL Server and that the MySQL Server firewall has been set up properly to accept connections from the reader's computer (see Chapter 11 for more information about cloud-based services).

Before connecting to an MySQL database, one must first install MySQL Open DataBase Connectivity (ODBC) drivers to allow Excel to connect. Then, one must set up a connection

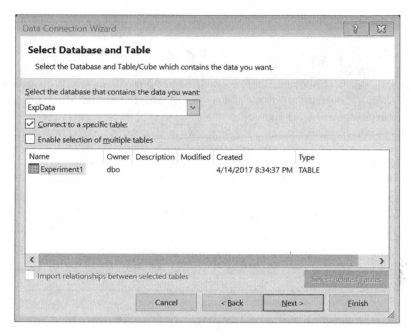

FIGURE 10.20

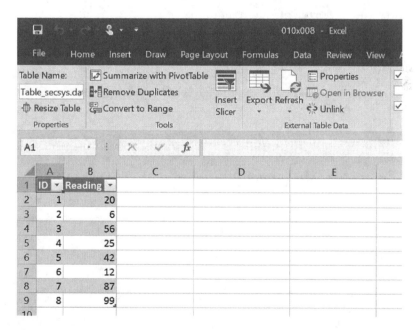

FIGURE 10.21

definition so that we can make a connection to the database server and pull data from a table in the database.

Information about downloading and installing the MySQL ODBC driver can be found on the Internet at https://dev.mysql.com/downloads/connector/odbc. This website provides excellent instructions on how to download and install the MySQL driver.

Once the driver is installed, an ODBC connection definition needs to be set up for the MySQL server using the newly installed driver. A simple search on Google or Bing using the term "ODBC setup" will provide ample instructions on how to configure the ODBC connection for your MySQL server.

Now that the MySQL ODBC connection is configured, it is time to configure Excel to connect to the MySQL server that holds the data in question. In this case, the data from Problem 3.1 are stored in a table named Experiment2 in an MySQL database named ExpData. To configure MS Excel to connect to this table, perform the following steps:

1. Open a new Excel workbook to Sheet1.
2. Click on DATA/GET EXTERNAL DATA/From Other Sources/From Data Connection Wizard.
3. A dialog box like that shown in Figure 10.22 will appear. Select ODBC DSN and click Next.
4. Click on the name of the ODBC DSN for the MySQL Server that was set up above. In the case of this example, the name is simply MySQL. Click Next once the name is selected.
5. The Data Connection Wizard dialog box will appear as in Figure 10.23. Because the MySQL ODBC connection already specified the database ExpData, the dialog box already has that database name indicated. As such, the dialog box displays the table name Experiment2.
6. Make sure the Experiment2 table name is selected and click Next; then, click Finish on the Save Data Connection File and Finish screen.
7. A dialog box similar to that in Figure 10.17 will appear. Click OK and the result will look similar to Figure 10.24.

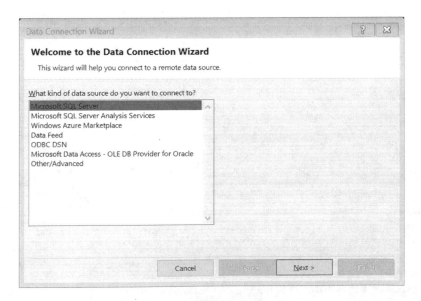

FIGURE 10.22

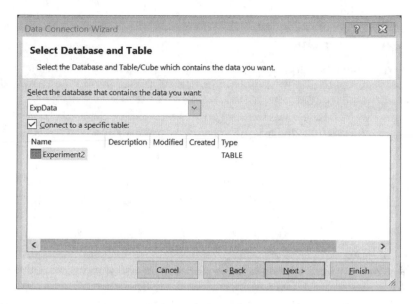

FIGURE 10.23

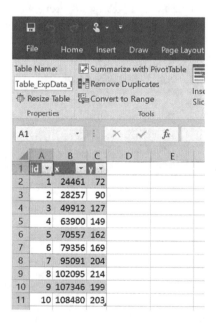

FIGURE 10.24

10.5.5 Connecting to Other Data Sources

Above, we have outlined connecting to several common data sources. Excel, however, supports many more options that will have built-in support (like MS Access or MS SQL Server), that will be accessible via an ODBC connection, or that will represent non-database sources (From Web, From Text, etc.). The reader is encouraged to explore these options for fun and gaining familiarity.

10.6 Microsoft Power Query

Microsoft Power Query is simply the DATA/GET & TRANFORM set of menu options in Excel 2016. Using this facility, we replicate obtaining badge swipe data from the Access database as outlined in Section 10.5.2.

Starting with the MS Access database file referenced in Section 10.5.2, we will get the data from the database using this alternate facility. To do so, perform the following steps:

1. Open a new Excel workbook.
2. Click on DATA/GET & TRANSFORM/New Query/From Database/From Microsoft Access Database. Notice the number and variety of standard database sources from which you can query.
3. A dialog box appears to select the appropriate MS Access database file from the file system. Find and select SecuritySystem.accdb as in the previous procedure. A navigator window will appear as in Figure 10.25. Select the BadgeSwipes object to view the table data. Click on Load.
4. The data will load into another worksheet as shown in Figure 10.26.

As with a Microsoft Access data source, Power Query can be used to get and transform data from a variety of sources: files, databases, Microsoft Azure, Online Sources, and Other data sources. The bulleted list below provides a listing of the categories and options available from the DATA/GET & TRANSFORM/New Query source options. The reader is encouraged to explore these options to gain experience and familiarity with their applicability.

FIGURE 10.25

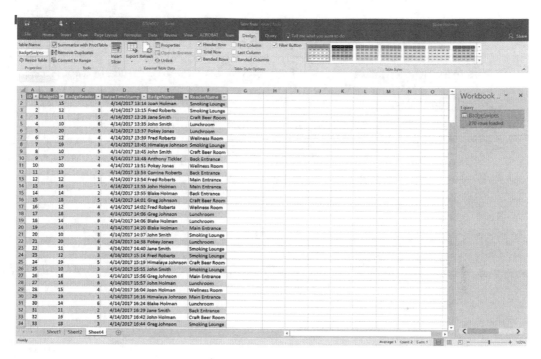

FIGURE 10.26

- Files: Workbook, CSV, XML, Text, JSON, Folder, MS SharePoint Folder
- Databases: MS SQL Server, MS Access, MS SQL Server Analysis Services, Oracle, IBM DB2, MySQL, PostgreSQL, Sybase, Terradata, SAP Hana
- MS Azure: MS Azure SQL Database, MS Azure SQL Data Warehouse, MS Azure Marketplace, MS Azure HDInsight, MS Azure Blob Storage, MS Azure Table Storage
- Online Services: MS SharePoint Online List, MS Exchange Online, MS Dynamics CRM Online, Facebook, Salesforce Objects, Salesforce Reports
- Other Sources: Web, MS SharePoint List, OData feed, Hadoop File, MS Active Directory, MS Exchange, ODBC, Blank Query

Problems

10.1 Consider the data set used in Example 10.1. Instead of constructing the Full Name to be First Name Middle Initial Last Name, modify the string concatenation formula to construct the Full Name as Last Name, First Name Middle Initial.

10.2 Consider the Total Number of Winners data available at the website www.powerball.com/powerball/pb_totalwinners.asp. Construct a worksheet in Excel that retrieves the data and inserts the data into a worksheet. For fun, using the techniques in Chapter 3, plot a vertical bar chart of total winners by US State.

10.3 Create an MS Access database with the data from Problem 3.1 in it. Create an Excel workbook that queries the data using Power Query.

10.4 Using the same Access database from Problem 10.3, create an Excel workbook that imports the data using the DATA/GET EXTERNAL DATA/From Access menu option. Comment on the difference between this method and that used in Problem 10.3.

10.5 Create a list of student IDs (column A) and student first names (column B) and student last names (column C) in a worksheet (Sheet1). In a separate worksheet (Sheet2), create a random list of student IDs in column A. In column B of Sheet2, create a formula using VLOOKUP that obtains the student first names and last names and concatenates them together in the format of last name, first name. Envision how this might work with rows and the HLOOKUP function instead.

11

Office 365 and Integration with Cloud Resources

11.1 Introduction

"The Cloud" is a term that has come about in the last 5 or so years to represent a broadening of the concept of service bureau computing that put Electronic Data Systems (EDS) on the map in the latter part of the 20th century. While service bureau computing focused on mainframe computing resources, Cloud computing is focused on several dimensions of computing for both commercial and consumer use, mostly in the context of midrange and minicomputer platforms. Cloud computing, sometimes referred to as grid computing, can involve many variations. For purposes of this text, we will focus on flexible computing services like Office 365, Microsoft Azure, and Amazon Web Services (AWS) as their business models allow customers to adjust their resource needs as demand fluctuates.

11.2 What Is Office 365?

Microsoft's Office 365 product is a collection of software capabilities ranging from simple use of the Microsoft Office suite of tools to an Enterprise set of tools that include corporate E-mail (MS Exchange Online), Social Enterprise (MS Yammer and Delve), Collaboration tools (MS SharePoint), and several others. Customers pay a monthly fee for access to the software and services, which include automated updates and feature enhancements. Full details can be obtained via Microsoft's website or a Microsoft account representative. Note that nonprofit and educational organizations can qualify for a significant price reduction due to their nonprofit or educational status.

11.3 Leveraging MS Excel on Premise and in the Cloud

Office 365 supports both a locally installed version of MS Excel as well as an online instance of MS Excel called, oddly enough, Excel Online. Imagine visiting a colleague and wanting to demonstrate a new whiz-bang Excel spreadsheet that you crafted that shows groundbreaking results from an experimental analysis. But, you don't have your computer with you and they don't have MS Excel (strange as it may sound). The advantage of Office 365 and Excel Online is that you can log into the Microsoft Office 365 portal and use Excel Online to view your spreadsheet and share the excitement with your colleague.

11.4 Integrating MS Excel with Cloud Resources

Thanks to the power of internetworking, Excel is no longer limited to data contained within an Excel spreadsheet or on a local or network disk drive close to your computer. With valid Internet connectivity, the possibilities are nearly endless for connecting to data and being able to manipulate it.

As demonstrated with several scenarios in Chapter 10, Excel can obtain data from other worksheets, workbooks, data files, databases, websites, and even online services like Facebook. Data are everywhere and Excel can get to the data either through the DATA/ GET EXTERNAL DATA menu options or the Power Query facility, noted as the DATA/ GET & TRANFORM menu options.

The reader is encouraged to explore the services that Excel supports to see how the data look when brought into an Excel workbook. Chapter 10 provides a great start to this exploration process. For example, Problem 10.2 points to data on the PowerBall website.

11.4.1 Microsoft Azure

In addition to Office 365, Microsoft has a cloud service offering for server-based computing capabilities. Called Microsoft Azure, it is a constantly evolving Microsoft service offering. More information can be found at the Microsoft Azure website at http://azure.microsoft. com. The reader is encouraged to consider signing up for a free account to evaluate and explore cloud-based Microsoft SQL Server database services on Azure.

11.4.2 Amazon Web Services

Similar to Microsoft, Amazon has a cloud service offering for server-based computing capabilities, offered under the AWS product family. AWS is a constantly evolving product offering from Amazon. More information about AWS can be found at http://aws.amazon.com. The reader is encouraged to consider signing up for a free tier account with AWS and explore cloud-based MySQL database services at AWS.

11.5 Excel and Microsoft Power BI

This entire text has focused on Excel as an excellent toolset for analyzing data, processing data, and charting the results of data analyses. Microsoft Power BI is a product that complements Excel with regard to charting data and presenting the data to users in a Dashboard format. Excel worksheets make an excellent data source for Microsoft Power BI as do database sources.

Figures 11.1 and 11.2 show sample data sources in Excel workbook form as inputs to Microsoft Power BI. They represent data regarding a cash room operation and show

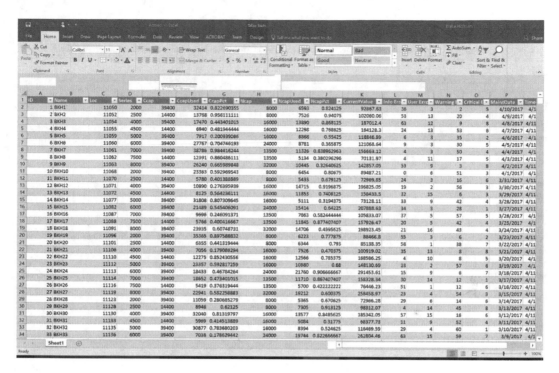

FIGURE 11.1

FIGURE 11.2

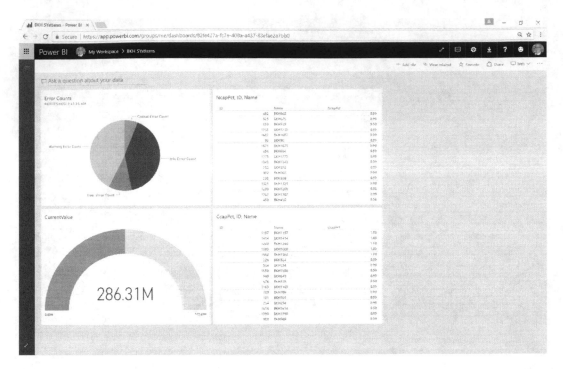

FIGURE 11.3

dashboard data regarding their operation. Figure 11.3 shows a sample dashboard created in Microsoft Power BI in under an hour, demonstrating the speed with which data dashboards can be created.

The reader is encouraged to sign up for a free account at www.powerbi.com and explore the use of this toolset in conjunction with Excel.

References

1. Hillier, F.S. and Lieberman, G.J., *Introduction to Mathematical Programming*, 2nd ed., McGraw-Hill, New York, 1995.
2. Holman, J.P., *Experimental Methods for Engineers*, 7th ed., McGraw-Hill, New York, 2000.
3. Holman, J.P., *Heat Transfer*, 9th ed., McGraw-Hill, New York, 2002.
4. Winston, W.L., *Operations Research, Applications, and Algorithms*, 3rd ed., PWS-Kent, Boston, MA, 1994.
5. Wu, N. and Coppins, R., *Linear Programming*, McGraw-Hill, New York, 1981.

Index